W9-ADV-936

How to Diagnose and Fix Everything Electronic

Michael Jay Geier

New York Chicago San Francisco Lisbon
London Madrid Mexico City Milan New Delhi
San Juan Seoul Singapore Sydney Toronto

The McGraw·Hill Companies

Library of Congress Cataloging-in-Publication Data

Geier, Michael (Michael Jay)
  How to diagnose and fix everything electronic / Michael Geier.—1
      p.   cm.
  Includes index.
  ISBN 978-0-07-174422-5 (pbk.)
  1. Electronic apparatus and applications—Maintenance and
repair—Handbooks,
manuals, etc.. I. Title.
  TK7870.2.G43   2011
  621.381028'8—dc22   2010052301

McGraw-Hill books are available at special quantity discounts to use as premiums and sales promotions, or for use in corporate training programs. To contact a representative, please e-mail us at bulksales@mcgraw-hill.com.

How to Diagnose and Fix Everything Electronic

2 3 4 5 6 7 8 9 0   DOC DOC   1 0 9 8 7 6 5 4 3 2 1

ISBN   978-0-07-174422-5
MHID   0-07-174422-3

Sponsoring Editor Roger Stewart	**Copy Editor** Lisa Theobald	**Composition** Glyph International
Editorial Supervisor Patty Mon	**Proofreader** Carol Shields	**Illustration** Glyph International
Project Manager Vasundhara Sawhney, Glyph International	**Indexer** Jack lewis	**Art Director, Cover** Jeff Weeks
Acquisitions Coordinator Joya Anthony	**Production Supervisor** George Anderson	**Cover Designer** Mary McKeon

This book is dedicated to my parents, for putting up with their young son's taking everything in the house apart, even though it scared them to death; to my brother, for providing me with a steady stream of broken items to fix and the encouragement to figure them out; to Greg, for sharing countless happy teenage hours fiddling with circuits, projects and walkie-talkies; to Rick, for always believing in and promoting my talents; to Cousin Jerry, for some of my earliest guidance in electronic exploration; and to Alvin Fernald and Tom Swift Jr., whose fictional technological exploits kept me spellbound through most of my childhood and made me believe anything was possible with a handful of transistors and the know-how to make them wake up and do something.

About the Author

Michael Jay Geier has been an electronics technician, designer and inventor since age 6. He took apart everything he could get his hands on, and soon discovered that learning to put it back together was even more fun. By age 8, he operated a neighborhood electronics repair service that was profiled in *The Miami News*. He went on to work in numerous service centers in Miami, Boston and Seattle, frequently serving as the "tough dog" tech who solved the cases other techs couldn't. At the same time, Michael was a pioneer in the field of augmentative communications systems, helping a noted Boston clinic develop computer speech systems for children with cerebral palsy. He also invented and sold an amateur radio device while writing and marketing software in the early years of personal computing.

Michael holds an FCC Extra-class amateur radio license. His involvement in ham radio led to his writing career, first with articles for ham radio magazines, and then with general technology features in *Electronic Engineering Times*, *Desktop Engineering*, *IEEE Spectrum*, and *The Envisioneering Newsletter*. His work on digital rights management has been cited in several patents. Michael has a Boston Conservatory of Music degree in composition, was trained as a conductor, and is an accomplished classical, jazz and pop pianist, and a published songwriter. Along with building and repairing electronic circuitry, he enjoys table tennis, restoring old mopeds, ice skating, bicycling, and banging out a jazz tune on the harpsichord in his kitchen.

Contents

Foreword

There is a keen sense of personal accomplishment to be gained from fixing something yourself that might otherwise have cost a fortune to repair or been recycled ahead of its time. Michael Jay Geier has known this joy since childhood. Now, for the first time, he shares his secrets in written and pictorial form.

I've known Michael across three decades, multiple time zones and dozens of entertainment and technology projects and consulting gigs we've tackled together. Quite simply, Michael sees electronic products as songs or symphonies of components, specialized parts working in harmony when they leave the manufacturer.

Yet individual musicians may be missing or off-key when the product fails from age, misuse or random component failure. Like the keen orchestra conductor he was trained to be, Michael quickly zeroes in on what parts of a broken electronic product are out of tune, using many skills he will teach you in this book, along with instruments to sense and measure things beyond the human senses of sight and sound. Fixing and extending the life of products we love, including things no longer being made and for which there is no ready replacement, is a valuable skill worth developing and nurturing.

Musicians see patterns well and communicate their art to larger audiences. Michael's expertise in troubleshooting consumer electronics is unmatched. Here, he shares the patterns that come easily to him with a broad audience of readers who want to enjoy their consumer electronics products longer, can't afford traditional repairs, fear their favorite irreplaceable gadget could be lost or further damaged while at the shop, or want to keep alive something old or obscure enough that no repair facility has the resources to work on it.

I put myself through college repairing consumer electronics and entertainment products. As much as I learned, when I first met Michael, I knew within days that he had a gift for troubleshooting far faster than my own. Jealousy soon faded as I saw he was confident and professional in his communications skills, and he loved sharing his insights and tricks.

Many TV, camcorder and video player manufacturers have employed lessons learned from Michael's shop repairs to improve their next-generation products and make them more durable and dependable. Michael has made his mark on improving product designs for more than two decades now.

I personally know several consumer electronics repair shop employers who hated to see Michael go. Yet not a one denied that Michael's time there and sharing of expertise made the entire shop better at repair and gave customers their serviced products back faster and with more reliability against ever failing again.

Herein, Michael shares his hunches, skills and insights at a level any dedicated reader can absorb and apply. Enjoy the satisfaction that comes with learning to repair your own equipment. And spread the word—it's about as green and economically smart as you can be!

Richard Doherty, Research Director
The Envisioneering Group
Seaford, New York
July 2010

Acknowledgments

To Neil Salkind, Roger Stewart, Joya Anthony and the other wonderful folks at McGraw-Hill, many thanks for recognizing the value of this material and shepherding it into existence. It takes a team to raise an idea. You've been a great team, and I'm honored to have been part of it.

Introduction

Everything. That's a scary word, one I almost avoided including in the title. Can any book actually cover *everything* about a topic? Yes and no. Yes, in the sense that the principles and techniques you'll learn can be applied to the repair of every kind of consumer electronics device presently being made or likely to be sold in the near future. No, in that it's impossible to fit each of the thousands of types of components and countless varieties of gadgets in the world into one volume. Covering all of them in deepest detail would take a library, and a good-sized one at that.

The focus of this book is on today's electronics, most of which are digital in nature, and the kinds of problems you're most likely to encounter. It might seem like there isn't that much one can service in modern digital gear, compared to the older analog circuitry. Dense boards populated by rows of chips with leads too close together even to poke at with a test probe don't seem like good repair candidates, do they? Luckily, those areas aren't where most failures occur, and there's still plenty of accessible circuitry to work on! In fact, some common problems in today's gear were rare or nonexistent in earlier technology, and they're quite reparable.

Exotic and very obsolete components and their associated products aren't covered in this book. Electron tubes, once the mainstay of all electronics, are pretty much gone, so we won't spend time on their peculiarities and specific troubleshooting methods. If you want to repair tube-type guitar amplifiers, you can find books dedicated to them. Similarly, we won't be discussing microwave ovens, which also have tubes, or transmitting amplifiers of the sort used by amateur radio operators. Nor will we take more than a passing glance at cathode ray tube (CRT, or picture tube)–based TVs and monitors. The CRT had a good long run, from the 1940s until just a few years ago, but it's a dead technology, thoroughly supplanted by flat-panel displays. Servicing CRT sets is rather dangerous, so please find a book devoted to them if you have an interest in, say, restoring antique TVs. What's covered here is relevant but not comprehensive enough regarding that topic to keep you safe around those high-voltage beasts.

Some obsolete technology is still in common use and may remain so for years to come, so we'll explore it. Tape-based video recording continues to be used in some digital camcorders. VCRs, which are rapidly disappearing as high-definition TV (HDTV) obsoletes them, may be the only key to recovery of precious home movies yet to be transferred to digital media. Serious audio devotees treasure their analog tape recorders and turntables and will never replace them with CD or MP3 players. We won't spend much time on the old formats, but the troubleshooting techniques covered here are applicable to their repair.

Most of today's digital equipment still contains analog circuitry for audio or video output, microphone input, voltage regulation and such. Home theater receivers use analog amplifier stages, and many have old-fashioned, linear power supplies as well, because they're electrically quieter than newer, pulse-driven designs. In fact, the best audiophile-grade stereo gear is pretty much all analog and will likely remain that way. Even digital radio and TV receivers use analog stages to amplify and separate incoming signals before digital decoders extract the data. So, troubleshooting techniques specific to analog circuitry are far from antiquated; they continue to be relevant in our digital era.

In this book, it is assumed that you have probably opened an electronic device at one time or another and checked a fuse. Perhaps you know a resistor when you see one, and maybe you've even soldered or done some basic troubleshooting. Still, we're going to start from the top, ensuring you're a sound swimmer before diving into the deep end. And dive we will! Beginning with a look at the tools you'll need, we'll explore setting up your home workshop. We'll discuss the best types of workbenches and lamps, and where to put your gear and tools. We'll take a close look at the most useful test instruments, where to find bargains on them, and how to operate them. Getting good with an oscilloscope is key to being a crack shot tech, so we'll explore a scope's operation in detail, button by button.

Using other test equipment like digital voltmeters and ohmmeters is also crucial to effective repair. We'll focus on commonly available test gear, without spending significant effort on very expensive, exotic instruments you're never likely to own. We'll examine how to take a product apart, figure out what's wrong with it, replace parts and close it back up again. Finally, we'll look at tips and tricks for specific devices, from optical disc players to video recorders and receivers. Here's a quick breakdown of what's in each chapter.

Chapter 1, "Prepare for Blastoff: Fixing Is Fun!"

- Why repair things? Environmental and economic factors, learning, fun, preserving rare and obsolete technology, potential profit.
- When is a product worth repairing, and when is it better to cut it up for parts?

Chapter 2, "Setting Up Shop: Tools of the Trade"

- Necessary items, from hand tools to test instruments, and how to buy them. Must-haves, nice-to-haves, and expensive goodies to dream about.
- How to select a workbench and set it up, and where to put it.

Chapter 3, "Danger, Danger! Staying Safe"

- How to avoid getting hurt while servicing electronics: electrical and physical hazards, eye and ear protection.
- How not to damage the device you're repairing: causing electrical and physical damage.
- Ensuring user safety after product repair.

Chapter 4, "I Fix, Therefore I Am: The Philosophy of Troubleshooting"

- General troubleshooting principles: why things work, why they stop.
- Common mistakes and how to avoid them.
- Organization of modern devices: microprocessor brains, nervous system, muscles and senses.
- The "art" side of electronics: manufacturer-specific quirks and issues.
- What fails most often and why.
- Failure history and how it helps diagnose problems.
- Preliminary diagnosis based on symptom analysis: dead, comatose and nearly working.
- Case histories.

Chapter 5, "Naming Names: Important Terms, Concepts and Building Blocks"

- Electrical units: volts, amps, resistance, capacitance, and so on.
- Circuit concepts: how parts connect and how current moves through them. Series and parallel.
- Signal concepts: how changes in voltage represent information. Waveforms. Analog and digital representation.
- Building blocks: common circuits used in many products. Amplifiers, oscillators, frequency synthesizers and power supplies.

Chapter 6, "Working Your Weapons: Using Test Equipment"

- Digital multimeter: measuring voltage, current and resistance.
- Oscilloscope: detailed, button-by-button operation, including delayed sweep measurements.
- AC and DC signal components, rolloff and other issues affecting measurement strategy.
- Soldering and desoldering techniques.
- Bench power supply: voltage and current considerations, DC plug polarity.
- Transistor tester.
- Capacitance meter.
- Signal generator.

- Frequency counter.
- Analog meter: when to use it, interpreting the wiggling meter needle, tests not possible with a digital instrument.
- Contact cleaner spray: what to use it on, what not to.
- Component cooler spray: solving thermal intermittents, considerations for safe use.

Chapter 7, "What Little Gizmos Are Made of: Components"

- Common parts: capacitors, clock oscillators, crystals, diodes, fuses, inductors and transformers, integrated circuits, op-amps, resistors, potentiometers, relays, switches, transistors, voltage regulators and zeners.
- Varieties of each type of part.
- Symbols, markings and photos.
- Uses: what components do in circuits.
- What kills them.
- How to test them out of circuit.

Chapter 8, "Roadmaps and Street Signs: Diagrams"

- Block, schematic and pictorial diagrams.
- Learning to read diagrams like a story: signal flow, organization in stages.
- Symbols and call numbers.
- Good, average and bad diagrams.
- Part-by-part analysis of individual stages and their functions. Amplifier example.
- Organization of larger structures. Switching power supply example.
- Practicing reading: looking for stages and structures in radios and DVD players.
- Working without a diagram.
- Case history of troubleshooting an LCD TV without a schematic.

Chapter 9, "Entering Without Breaking: Getting Inside"

- Separating case halves: hidden snaps.
- Disconnecting ribbon cables.
- Layers: disassembling in order, use of digital photos and nested cups.
- Disassembly tips for common products: receivers, VCRs, DVD players, flat-panel TVs, turntables, video projectors, MP3 players, PDAs, cell phones, camcorders, digital cameras and laptop computers.

Chapter 10, "What the Heck Is That: Recognizing Major Features"

- What various sections of circuitry look like: descriptions and photos.
- Recognizing sections from components specific to their functions: inductors, power transistors, and so on.
- Power supplies: linear and switching.
- Backlight inverters.

- Signal processing areas, analog and digital.
- Digital control sections.
- Output stages: discrete transistors and integrated modules.
- Mechanisms: video head drum, capstan motor, laser optical head and DLP color wheel.
- Danger points.

Chapter 11, "A-Hunting We Will Go: Signal Tracing and Diagnosis"

- Where to begin, based on observed symptoms.
- Dead, comatose or crazy, alive and awake but not quite kicking.
- Intermittents: thermal and mechanical, bad solder joints, board cracks, positional and vibration-sensitive.
- Working forward or backward through stages: when each technique is appropriate.
- Stages, test points and making sure you're in the right place.
- Zeroing in on bad components.
- Desperate measures: shotgunning, current blasting and LAP method.

Chapter 12, "Presto Change-O: Circuit Boards and Replacing Components"

- Desoldering through-hole and surface-mount components.
- Choosing replacement parts: new, from your stash and from parts machines.
- Substituting similar parts when you can't get the exact replacement: vital characteristics that must be matched or exceeded, and allowable differences in capacitors, diodes, resistors, transistors and zeners.
- Installing new parts: through-hole and surface-mount, mounting power transistors.
- Finding components: standardized, proprietary, local, mail-order, new and surplus.
- Saving damaged boards: bridging broken conductors and bad layer interconnects.
- Reflowing solder on high-density integrated circuit chips.

Chapter 13, "That's a Wrap: Reverse-Order Reassembly"

- Common reassembly errors.
- Ensuring good ground connections on boards and chassis.
- Lead dress: placement of wires and cables, physical and thermal risks, electromagnetic interference.
- Reconnecting ribbon cables.
- Repairing damaged ribbon sockets.
- Reversing layer and cup order.
- Rejoining plastic snaps.
- Reinserting screws: tension and correct placement.
- Final test.

Chapter 14, "Aces Up Your Sleeve: Tips and Tricks for Specific Products"

- How they work, what can go wrong, when repair is worth doing, dangers within, and how to fix them.
- Switching power supplies, receivers, disc players and recorders, flat-panel displays, hard drives, laptop computers, MP3 players, VCRs, camcorders and video projectors.

Whether or not you've already had your hands inside some electronic devices, this book will guide you from the "maybe it's the fuse" level to the "ah, the biasing diode on the output stage is open" point. It will help hone your sleuthing skills with logic and a solid foundation in how things work, until you feel like an ace detective of electrons. At the very least, it'll leave you fascinated with everything that goes on inside your favorite gadgets and eager to tackle everything that comes your way. *Everything*...maybe it's not such a scary word after all.

Chapter 1

Prepare for Blastoff:
Fixing Is Fun!

Electronics is a lifelong love affair. Once its mysteries and thrills get in your blood, they never leave you. I became fascinated with circuits and gadgets when I was about 5 years old, not long after I started playing the piano. There may have been something of a connection between the two interests—both involved inanimate objects springing to life by the guidance of my mind and hands. Building and repairing radios, amplifiers and record players always felt a little like playing God, or perhaps Dr. Frankenstein: "Live, I command thee!" A yank on the switch, just like in the movies, and, if I had figured out the puzzle correctly (which was far from certain at that age), live it would! Pilot lights would glow, speakers would crackle with music and faraway voices, and motors would turn, spinning records that filled my room with Haydn, Berlioz and The Beatles. It was quite a power trip (okay, a little pun intended) for a kid and kept me hankering for more such adventures.

By age 8, I was running my own neighborhood fix-it business, documented in an article by *The Miami News* titled "Little Engineer Keeps Plugging Toward Goal." Repairs usually ran about 25 cents, and I had customers! Neighborhood pals, their families and my dad's insurance business clients kept me busy with malfunctioning radios and tape recorders. I even fixed my pediatrician's hearing tester for 50 cents. If only I'd known what *he* was charging....

My progression from such intuitive tinkering to the understanding required for serious technician work at the employable level involved many years of hands-on learning, poking around and deducing which components did what, and tracing signals through radio stages by touching solder joints with a screwdriver while listening for the crackling it caused in the speaker. Later came meters, signal tracers and, finally, the eye-opening magic window of the oscilloscope.

Ah, how I treasure all the hours spent building useful devices like intercoms and fanciful ones like the Electroquadrostatic Litholator (don't ask), fixing every broken gadget I could get my hands on, and devouring *Popular Electronics*, *Electronics Illustrated* and *Radio-Electronics*—great magazines crammed with construction articles and repair advice columns. Only one issue a month? What were they waiting for??

C'mon, guys, I just have to see the last part of that series on building your own color TV camera, even though I'll never attempt it. But now I know how a vidicon tube works! And, thanks to my parents' wise and strict rule that I experiment only on battery-powered items, I survived my early years to share my enthusiastically earned expertise with you, the budding tech.

After graduating from the Boston Conservatory of Music, I did what any highly trained, newly certified composer/conductor does: I completely abandoned my field of study and started working in electronics! I was a tech in repair shops, I programmed computers, and I developed circuitry and software for several companies around Boston and New York, while building my own inventions and running a little mail-order company to sell them. All of those experiences integrated into the approach I will present in this book, which includes inductive and deductive reasoning, concepts of signal flow and device organization, taking measurements, practical skills and tips for successful repair, a little bit of art, and even a touch of whimsy here and there.

No book can make you an expert at anything; that takes years of experience and squirreling away countless nuggets of wisdom gleaned from what did and didn't work for you. My hope is that this distillation of my own hard-won understanding will infect you with the love of circuits and their sometimes odd behaviors, and start you on the very enjoyable path of developing your skills at the wonderful, wacky world of electronic repair.

So, warm up your soldering iron, wrap your fingers around the knobs of that oscilloscope and crank up the sweep rate, 'cause here we go!

Repair: Why Do It?

When I was a kid, there were radio and TV servicers in many neighborhoods. If something broke, you dropped it off at your local electronics repair shop, which was as much a part of ordinary life as the corner automotive service garage. These days, those shops have all but disappeared as rising labor costs and device complexity have driven consumer electronics into the age of the disposable machine. When it stops working, you toss it out and get a new one. So why fix something yourself? Isn't it cheaper and easier just to go out to your local discount store and plunk down the ol' credit card?

It might be easier, but it's usually not cheaper! Sans the cost of labor, repair can be quite cost effective. There are lots of other good reasons to become a proficient technician, too:

- *It's fun.* You'll get a strong sense of satisfaction when your efforts yield a properly working gadget. It feels a bit like you're a detective solving a murder case, and it's more fun to use your noodle than your wallet.
- *It's absorbing.* Learning to repair things is a great hobby to which you can devote many fruitful hours. It's good for your brain, and it beats watching TV any day (unless you fixed that TV yourself!).

- *It's economical.* Why pay retail for new electronics when you can get great stuff cheap or even free? Especially if you live in or near a city, resources like craigslist.org will provide all the tech toys you want, often for nothing. Lots of broken gadgets are given away, since bringing them in for repair costs so much. They're yours for the taking. All you have to do is fix 'em!
- *It can be profitable.* Some of the broken items people nonchalantly discard are surprisingly valuable. When your tech skills become well developed, you'll be able to repair a wide variety of devices and sell what you don't want for yourself.
- *It can preserve rare or obsolete technology.* Obsolete isn't always a negative term! Some older technologies were quite nice and have not been replaced by newer devices offering the same features, utility or quality. The continued zeal of analog audio devotees painstakingly tweaking their turntables offers a prime example of the enduring value of a technology no longer widely available.
- *It's green.* Every product kept out of the landfill is worth two in ecological terms: the one that doesn't get thrown away and the one that isn't purchased to replace it. The wastefulness of tossing out, say, a video projector with a single capacitor is staggering. To rip off an old song, "Nothing saves the green'ry like repairing the machin'ry in the morning...."
- *Your friends and family will drive you crazy.* Being a good tech is like being a doctor: everyone will come to you for advice and help. Okay, maybe this one isn't such an incentive, but it feels great to be able to help your friends and loved ones, doesn't it? Being admired as an expert isn't such a terrible thing either.

Is It Always Worth It?

While it's often sensible to repair malfunctioning machines, sometimes the endeavor can be a big waste of time and effort, either because the device is so damaged that any repair attempt will be futile or the cost or time required is overwhelming. Part of a technician's expertise, like a doctor's, lies in recognizing when the patient can be saved and when it's time for last rites and pulling the plug—in this case, literally! Luckily, in our silicon and copper realm, those destined for the hereafter can be recycled as parts. A stack of old circuit boards loaded with capacitors, transistors, connectors and other components is as essential as your soldering iron, and you'll amass a collection before you know it.

Chapter 2

Setting Up Shop: Tools of the Trade

To repair anything, you will need some basic test gear and a suitable place to use it. Because electrons and their energy flows are invisible, test equipment has been around almost as long as human awareness of electricity itself. The right test instruments and hand tools enable you to get inside a product without damaging it, find the trouble, change the bad parts and reassemble the case correctly and safely.

Must Haves

Electronics work can involve a seemingly unending array of instruments, but you don't need them all. Some of them are insanely expensive and only rarely useful. Others cost a lot less and find application in almost every circumstance. Some items are absolutely essential, so let's start by looking at those things you can't live without, and how and where to set them up for the most effective, efficient service environment.

A Good Place to Work

Like surgery, tech work is exacting; there's little room for error. One slip of the test probe can cause a momentary short that does damage worse than the problem you were trying to solve. One of the most important elements of effective, conscientious repairing is an appropriate workspace set up to make the task as easy and comfortable as possible, minimizing the likelihood of catastrophic error.

First, consider your location. If you have young children, it's imperative that the workbench is set up in a room that can be locked. Opened electronic products and the equipment used to service them are not child-safe, and the last thing you or your kids need is an accident that could injure them. Dens and basements can be suitable locations, but garages are probably best avoided if the kids are still at that "poke in a finger and see what happens" age. Pets, too, can wreak all kinds of havoc on disassembled machinery. Cats love to climb on and play with things, particularly if those things are warm. The effects can range from lost screws and broken parts to dead cats! Let's face it, cats are not big readers, and a "Danger! High Voltage!" sticker looks about the same to them as "Cat Toy Inside." Keep kitty away from your repair work, even when you're in the room. You just never know when the little angel sitting there so placidly will make a sudden leap at your project and turn it into op-art.

Many of us have our workshops in the basement. This location is a mixed bag. It keeps the somewhat messy business of repair out of your living space, but it has some drawbacks. If you live in a cooler clime, it can get mighty chilly down there in the wintertime! Worse, basements tend to be damp, which is bad for your test gear. In damp environments, oscilloscopes and meters have a way of not working if you haven't used them for awhile, because moisture gets into connectors and redirects normal current paths in unpredictable ways. Still, the basement may be your best bet. Just be sure to fire up your gear now and then to dry it out, and run a dehumidifier if humidity climbs above 70 percent or so. Use an electric heater in the winter; kerosene heaters designed for indoor operation still emit quite a bit of carbon dioxide that will build up in the unventilated spaces of most basements. And should such a heater malfunction and put out a little carbon monoxide, you'll probably be dead before becoming aware of anything wrong.

The workbench itself should be as large as you can manage, with plenty of space for your test equipment, soldering iron, power supplies and other ancillary gear along the edges; you'll need to keep the center clear for the item to be repaired. Wonderful, prefab test benches can be mail-ordered, but they're fairly expensive and are most often found in professional shops. If you have the means, go for it. Get one with shelves and lots of power strips. If, like most of us, you'd rather not spend hundreds of dollars on a bench, there are plenty of alternatives. You can make your own in the time-honored way, from an old solid door (hollow doors aren't strong enough) and some homemade wooden legs and braces. If you're not the woodworking type, a big desk can sometimes suffice.

Sturdy desks and tables suitable as workbenches can often be had for very little from thrift stores or for free from online trading boards, because of one factor in your favor: they don't have to be pretty. In fact, avoid spotless, fancy furniture, because you'll feel bad when you nick, scrape, singe and accidentally drill holes in it. An Ikea-style desk works great, as long as it's well-braced and sturdy. A white Formica surface is nice too, because you can see dropped screws and such much more easily than with a darker, textured covering (see Figure 2-1). Don't even consider covering the bench with carpeting; you'll lose so many parts in it that you could eventually shake it out and build a fusion reactor from what you find! Also, carpeting can build up static electric charges lethal to circuitry.

Carpeting on the ground around the bench has its pros and cons. It's easy to lose small parts in it, but it also helps prevent them from bouncing away into oblivion when

FIGURE 2-1   The well-appointed workbench:
a little smaller than ideal, but it does the job

they fall. If you do choose to have a carpeted floor, pick a light color and as shallow and tight a pile as possible. This is no place for a thick carpet with loose fibers.

You'll need modern, three-wire (grounded) electrical service at your bench. This is critical for safety! A grounding adapter plugged into a 1920s two-wire outlet will not do, even if you screw the adapter's ground lug to the wall plate. Most of those plates are not properly grounded, and a bad ground can get you *killed* in certain circumstances.

The current (amperage) requirement is not high for most service work. Your scope and other instruments won't eat a lot of power, and most benches can be run quite safely using a single, modern 15-amp plug fanned out by a couple of hardware store-variety power strips. Also, this arrangement has the advantage that all ground points are at exactly the same voltage level, which helps prevent ground loops (unwanted current between ground points). Again, be sure the strips are three-wire, grounded types.

Lighting is another very important factor that shouldn't be ignored. While it might seem obvious that the entire room should be brightly lit, that is not the most productive approach, as it can actually make it harder to see small details that need to be scrutinized and, therefore, brighter than their surroundings. Average lighting in the room is adequate. What you need most is spot lighting, and the best solution is a fluorescent light on a swing arm, as shown in Figure 2-1. If it has a magnifier, all the better, but you'll be wearing one anyway, so it's not necessary.

Forget about using an incandescent bulb; the heat it produces will cook your hands, your face and the gadget you're trying to fix. An inexpensive way to obtain the necessary lighting is to get a swing-arm desk lamp and replace its incandescent bulb with an "eco bulb," one of those now-ubiquitous spiral light bulb replacements. Be aware, though, that many eco bulbs have a rather yellowish tint and also put out a fair amount of ultraviolet light, so using one close to your eyes may not be comfortable. Plus, they operate at a high frequency and can emit significant short-range radio-frequency energy capable of interfering with some kinds of measurements or even the circuit under test. The old circular, bluish-white fluorescent lamp is still your best bet.

Digital Multimeter

A *multimeter* (pronounced "mull-TIH-mih-ter") is a device that can test several electrical parameters. The most common and important quantities you'll need to measure are voltage (volts), resistance (ohms) and current (amps or, more typically, milliamps, which are thousandths of an amp). The analog incarnation of this test device, recognizable by its big meter needle and multiple-stop selector knob, used to be called a VOM (volt-ohm-milliammeter). Now that the meters are digital, they're usually called DMMs (Digital Multi Meters), but they do the same thing, except that the readout is numerical instead of something interpreted from the position of a meter needle.

DMMs began as very expensive, high-end laboratory instruments, but they're cheap now and pretty much all you can buy. The market positions have reversed, and VOMs have become the exotic technology, with a good one selling for considerably more than a digital. Hardware stores and RadioShack (a.k.a. TheShack) offer DMMs for around $20 to $50, and they're on sale on occasion for as little as $5. Some, however, can still be in the range of $200 or more. The expensive ones may have the ability to test various other parameters like capacitance and inductance, but mostly what they offer are much higher precision and accuracy.

Precision and accuracy are two different things. Precision is the fineness to which a measurement is specified, and accuracy is how truthful the measurement is. For instance, if I say, "It's between 60 and 80 degrees outside," and the actual temperature is 72 degrees, my statement is not very precise, but it's quite accurate. If, however, I say, "It's 78.69 degrees outside," and it's really 82 degrees, my statement is very precise but not at all accurate.

So, for a DMM to specify that it measures voltages to three digits to the right of the decimal point, it has to have a basic accuracy of somewhere around a thousandth of a volt. Otherwise, those pretty digits won't mean much! Who on Earth would build an instrument that displayed meaningless numbers?

Makers of low-cost DMMs do it all the time. The digits make one manufacturer's unit look more desirable than another's, but the basic accuracy doesn't support them. Does it matter? Not really, as long as you are aware of the limitations of the instrument's basic accuracy, so you know what to ignore toward the right side of the display. In any event, all DMMs, even the cheapies, are both more precise and more accurate than any VOM ever was.

Just how much precision do we need? For general service work, not a lot. When things break, they don't do so in subtle ways. For example, if you're checking the output of a 5-volt power supply, and your DMM reads 5.126 volts, that's not cause for concern. If it reads 3.5 volts or 7 volts, then perhaps you've found a problem! Bottom line: you don't need a $200 DMM. The $20 to $50 instruments will do fine.

Oscilloscope

Many hobbyists feel intimidated by the oscilloscope, but it is the best buddy any tech can have. Repeat after me: "My scope is my friend." Come on, say it like you mean it! Once you get the hang of using one, you will love it, I assure you.

The basic function of an oscilloscope is to generate a graph of voltage versus time. As a spot sweeps from left to right across the screen at a constant rate, it also moves up and down in relation to the incoming signal voltage, drawing a *waveform*, or representation of the signal that shows you how the voltage is changing. The maximum rate of change of the voltage amplifiers driving the vertical motion determines the bandwidth, or how fast a signal the scope can display. Most scopes in our range of interest have bandwidths of around 100 MHz (100 million cycles per second). They also have two *vertical input* channels, meaning they can display two waveforms at the same time. In Chapter 6, we'll explore how to use a scope—it really isn't hard—but first you have to get one. There are several types.

Analog

This is the classic scope with a green CRT (cathode ray tube, a.k.a. picture tube). It displays signals as they arrive and has no memory functions to store waveforms. It doesn't sample them, it doesn't dice or slice them; it just shows them to you, plain and simple. A classic analog scope is shown in Figure 2-2.

Analog scopes have been available since around the 1940s, and they really got good in the 1970s. Some are still being made today, though digital scopes have been at the forefront of the marketplace for a decade or more. Newer is better, right? Not always. The oscilloscope is a good example of an older analog technology being superior in some ways to its replacement. For most general service work, an analog scope is the simplest to use, and its display is the easiest to interpret. Further, it shows details of the signal that digital scopes may miss.

The lowest-end analog scopes have just one channel of input, and they lack features like delayed sweep, a very handy function that lets you zoom in on any part of a waveform you want and expand it for detailed viewing. Avoid them. There are tons of great analog scopes with all the nice features on the used market at ridiculously cheap prices, so there's no need to skimp on the goodies. Make sure any analog scope you buy has two channels (some have even more, but two are standard) and delayed sweep. Look for two input connectors marked "A" and "B" or sometimes "Channel 1" and "Channel 2," indicating two vertical input channels. If you see a metal knob that can be turned multiple times and slowly advances a number imprinted on it, or you see "A and B" on the big knob marked "Sec/Div" or "Horiz Sweep," or buttons marked "B after A," "B ends A," or "B delayed," then the unit has delayed sweep. Another tipoff

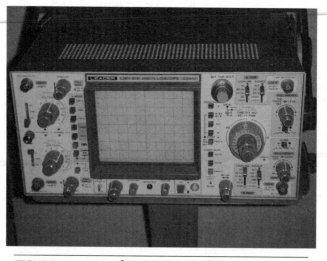

FIGURE 2-2 Leader LBO-518 100-MHz analog oscilloscope

is a knob marked "Trace Sep." If you're still not sure, look up the model on the Internet, and you can probably find its specs or even download a free PDF of the manual.

Digital

Digital scopes are new enough that they all have delayed sweep and two channels, except for a few handheld models. Though some early examples used CRTs, modern digital scopes can be recognized by their shallow cases and LCD (liquid crystal display) screens (see Figure 2-3). With a digital instrument, you can grab a waveform and examine it in detail long after it has ceased. Thus, digital scopes are ideal for working on devices with fleeting signals you need to be able to snag that may zip by only once.

Such is almost never the case in the kind of service work you'll be doing. The vast majority of the time, you will be looking at repetitive signals that don't have to be stored, and the limitations of a digital scope may get in your way.

One significant limitation arises from the basic nature of digital *sampling*, or digitizing, in which a voltage is sampled, or measured, millions of times per second, and the measured value of each sample is then plotted as a point on the screen. Alas, real-life signals don't freeze between samples, so digital scopes miss some signal details, which can result in a phenomenon called *aliasing*, in which a signal may be seriously misrepresented. (This is the same effect that causes wagon wheels in old Westerns to appear to rotate backward—the movie camera is missing some of the

FIGURE 2-3   Tektronix TDS-220 100 MHz digital oscilloscope

wheels' motion between frames.) If the sampling rate is considerably faster than the rate of change of the signal being sampled, aliasing won't occur. The sampling rate goes down, though, as you slow the *sweep rate* (speed of horizontal motion on the screen) down to compress the graph and squeeze more of the signal on the screen. As a result, when using a digital scope, you must always keep in mind that what you're seeing might be a lie, and you find yourself turning the sweep rate up and then back down, looking for changes in the waveform suggestive of aliasing. It takes some experience to be certain what's on the screen is a true representation of the signal. Even so, sometimes aliasing is unavoidable at lower sweep rates, limiting how much of the signal you can view at once—a conundrum that never occurs with analog scopes.

Another big limitation arises from the screen itself, and it also limits how much you can see at one time. Unlike the continuously moving beam of the analog scope, the digital scope's display is made up of dots, so it has a fixed *resolution*, and nothing can be shown between those dots. (That's why the sampling rate goes down at lower sweep rates; there's no point in taking samples between dots, since there's no place to plot them anyway.) When examining complex waveforms like analog video signals, the result is a blurry mess unless you turn the sweep rate way up and look at only a small part of the signal. While an analog scope can show a useful, clear representation of an entire field of video, a digital instrument simply can't; all you see is an unrecognizable blob.

Probably the most profound difference between an analog and a digital scope is that an analog instrument actually writes the screen at the sweep speed you select, while a digital unit does not. A digital collects the data at that speed, but it updates the screen much slower because LCDs don't respond very fast. For many signals, that's fine, and it can even help you see some signal features that might be blurred by repetitive overwrites on an analog screen.

Sometimes, however, those overwrites are exactly what you want. When viewing the radio-frequency waveforms coming from video and laser heads, for instance, you need to evaluate the *envelope*, or overall shape of the waveform over many cycles, not its individual waves. The overwriting and true-to-life writing speeds inherent in an analog scope make envelopes stand out clearly. Some envelopes can't be viewed *at all* with a digital, because it misses too much between screen updates. You'll see individual cycles of the waveform, but not their outer contour, unless you slow down the sweep rate so low that all you get is a featureless blur.

On the plus side, digital scopes are naturals at measuring waveforms, not just displaying them. They can provide a numerical readout of peak voltage, frequency, time difference, phase angle, you name it. While all of those measurements can be done with an analog scope, the computer in that case is your brain; you have to do the math, based on what you're seeing on the screen. With a digital scope, you position cursors on the displayed waveform and the scope does the work for you. Having quick and easy measurement of signal characteristics can greatly speed up troubleshooting.

When choosing a digital scope, look for the sample rate and compare it to the vertical bandwidth. The sample rate should always be higher than the bandwidth so the scope can perform *real-time sampling*. The Tektronix TDS-220, for example (Figure 2-3), samples at 1 gigasample (billion samples) per second, with a bandwidth of 100 MHz. Thus, one cycle of the fastest waveform it can display will be broken into ten samples, which is pretty good. At a minimum, the sample rate should be four times the bandwidth.

It is possible to sample repetitive waveforms at a rate slower than the bandwidth using a technique called *equivalent-time sampling*, in which each successive waveform is sampled at different points until the full representation is assembled. Equivalent-time sampling was developed when analog-to-digital converters were too pokey for real-time sampling of fast waveforms. It is an inferior technique, because developing an accurate representation requires the incoming signal to remain unchanged from cycle to cycle for as many cycles as it takes to assemble one. Plus, what you see is never a true picture of any one particular cycle. And, heck, it's just plain slow. Avoid any scope depending on it to reach its bandwidth specs. Real-time sampling is the only way to fly.

Analog with Cursor Measurement

This is the best of both worlds: an analog scope capable of performing many of the measurements available in a digital instrument. This style of scope doesn't digitize signals, thus avoiding all of the limitations associated with that process. Like a digital, though, it uses movable cursors to mark spots on the displayed waveform and calculate measurements. This is my favorite type of scope.

Analog with Storage

Before the advent of digital scopes, some analog units were made with special CRTs that could freeze the displayed waveform, enabling a crude form of signal storage. These scopes were expensive and always considered somewhat exotic. Digital storage has completely supplanted them.

PC-Based

The PC-based scope uses your general-purpose computer as a display and control system for a digitizing scope. It seems like a great idea, because you get a nice, big, high-resolution screen, and you can use your computer's keyboard and mouse to control the features. Plus, PC scopes are cheaper, since you're not paying for knobs and an LCD. In practice, PC scopes are the worst option for service work. They're awkward to use and usually offer the lowest performance in terms of sampling rate. I recommend you avoid them.

Buying an Oscilloscope

New scopes are fairly expensive. Expect to pay around $1000 for a 100-MHz instrument. But why spend a lot when there are so many nice scopes on the used market for next to nothing? It's quite possible to get a good used scope for about $100.

There are plenty of scope manufacturers, but the gold standard in the oscilloscope world is Tektronix. Tek has dominated the scope market since the 1970s, and for good reason. Many of its model 465 and 475 scopes from that era are still going strong, more than 30 years later! If you find one of those models in good working order for less than $100, it's worthy of consideration. Much newer models are also available, including the 2200 and 2400 series, and they're pretty cheap too. Other good scopes are made by Hitachi, Hewlett-Packard, B&K Precision and Leader. Fluke, famous for its DMMs, makes a series of handheld digital scopes, too, as does Tek. They're a tad pricey, though, and a bit harder to use, thanks to their extensive use of menus, since they have little room for knobs.

Where do you find a used oscilloscope? Good old eBay is loaded with them, and they show up now and then on craigslist.org. Try going to your area's hamfest, a periodic swap meet put on by ham radio operators and electronics aficionados, and you'll see plenty of scopes. Just be sure you can check that the instrument works properly before you plunk down your cash. Read Chapter 6 first so you'll know how to test the scope. Look for a nice, sharp trace and no lines burned into the display tube, if the scope uses one.

Along with the scope, you'll need a pair of probes. Scope probes are more than just pieces of wire; they have voltage dividers in them and are specialized devices designed to permit accurate signal measurement. Most divide the incoming voltage by 10 (you'll see why later) and are called 10X probes. Some have switches to remove the division, and are known as switchable 10X/1X probes (see Figure 2-4). Like scopes themselves, probes are rated by bandwidth, and the high-end ones can cost a lot. Luckily, 100-MHz probes can be found on eBay brand new for around $15 each. New probes will include slip-on covers with handy hooks on the ends. If you buy used probes, try to get the hooks too.

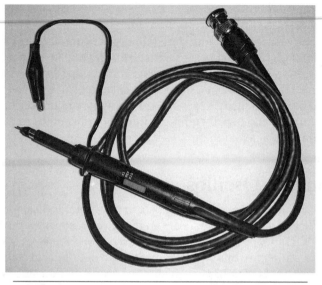

FIGURE 2-4   Oscilloscope probe, 10X/1X
switchable

Soldering Iron

Soldering irons come in various shapes and sizes, and you'll probably wind up with
more than one. The smallest, with heating elements in the 15-watt range, are great
for getting into very tight spots and working on tiny surface-mount parts, at least on
boards assembled with low-temperature solder. Those irons don't generate enough
heat to solder a power transistor, though. The largest irons, usually pistol-shaped
guns with elements of 100 watts or more, put out lots of heat and have sizable tips to
transfer it to the part you're soldering. Those big guns can be real life-savers, but you
sure don't want to try soldering minuscule parts with them. Even if you could fit the
tip where you needed it, the excessive heat would destroy the part and probably the
circuit board as well.

The best choice for general soldering work on printed circuit boards is an iron
with a medium-sized tip and a heating element in the range of 40 to 70 watts. Melting
leaded solder requires a tip temperature of about 375–400 °F. The newer, lead-free
variety needs a much hotter tip, in the area of 675–700 °F. Some inexpensive irons
in the 20-watt range are about the same size, but steer clear of those. Supplying
inadequate heat can cause lots of harm; you may easily pull up copper traces and
severely damage the printed circuit board if things aren't hot enough, especially when
removing components. Plus, not using enough heat can result in "cold" solder joints
that don't transfer electrical energy properly, causing your repair to fail.

Many inexpensive irons of medium size plug directly into the wall. This is not
the best way to go, as it may expose the circuitry you're soldering to small leakage

currents from the AC line, and having the cord go off to a power strip can be awkward as you move the iron around. Finally, should you accidentally lay the iron on the cord and melt through the insulation, you'll cause a short directly across the AC line, which is likely to be spectacular and unpleasant, and possibly dangerous. Don't laugh, it happens!

A far better solution is an iron that plugs into a base unit with a step-down transformer. This kind of setup runs the heating element at low voltage and isolates the tip from the AC line. The base unit gives you a nice stand to hold the iron and a sponge for wiping the tip, too. Some bases have variable heat controls, and some even have digital temperature readouts. Before the age of lead-free solder, I never found such things to be useful, because the heat pretty much always needed to be turned all the way up on smaller irons anyway. These days, a variable-heat soldering station that can hit the temperatures required for lead-free soldering is well worthy of consideration.

Numerous companies make soldering irons, but two make the nicest, most durable irons, the ones found in service shops: Weller and Ungar. These irons can cost from $50 to more than $100, but they are worth every penny and will last for many years. Your soldering iron is usually the first thing you turn on and the last you turn off, so it will run for thousands of hours and needs to be well-made. Don't be tempted by those $20 base-unit irons flooding the hobbyist market. They just don't hold up, and you'll be needing a new one before you know it.

As with scopes, good used irons often show up at great prices at hamfests. Wherever you get your iron, plan on buying a spare tip or two. Tips wear out and become pitted and tarnished to the point that they no longer transfer heat well, so they must be replaced every few years. The heating elements can go bad too, but it's rare; I've seen them last for decades on the good irons.

The big guns are cheap, typically under $20, so buy one. There will be situations in which you will be very glad you did.

Plastic-Melting Iron

Sooner or later, you'll want to melt some plastic to repair a crack or a broken post. It's unhealthy to breathe in molten plastic fumes, but we all melt the stuff now and then, being as careful as we can with ventilation. If you're going to melt plastic, don't do it with the same iron you use for soldering! The plastic will contaminate and pit the tip, making it very hard to coat it with solder, or *tin* it, for subsequent soldering work. Instead, pick up a cheap iron in the 20- to 30-watt range and dedicate it for plastic use. For this one, you don't need a base unit or any other fancy accoutrements. You should be able to get a basic iron and a stand to keep it from burning its surroundings for around $10.

Solder

Traditional solder is an alloy of tin and lead with a rosin core that facilitates the molecular bond required for a proper solder joint. In the past, the alloy was 60 percent tin, 40 percent lead. More recently, it has shifted to 65 percent tin, 35 percent lead. This newer type of solder is better suited to the lower temperatures associated with tiny surface-mount parts, and it's getting hard to find the old 60/40 stuff anymore. The old proportions were better for the higher-heat environment of power transistors and voltage regulators. If you can find some 60/40, it's worth getting. If not, you can live with the newer variety. Lead is a toxic metal, and lead-free solder has become available and is widely used in the manufacture of new electronics, to comply with the legal requirements of some countries and states. The European "Restriction of Hazardous Substances" or RoHS standard is gradually being adopted around the world. All products displaying the RoHS mark are made with lead-free solder. You can buy the stuff for your repair work, but I recommend against doing so, because it's hard to make good joints with it. It doesn't flow well, and cold joints often result. Plus, the higher heat required to melt it invites damage to the components you're installing.

Lead vaporizes at a much higher temperature than that used for soldering. The smoke coming off solder is from the rosin and does not contain lead you could inhale. Handling solder, however, does rub some lead onto your hands. So never snack or touch food while soldering, and always wash your hands thoroughly after your repair session ends.

There is a variety of solder, found in hardware stores and intended for plumbing applications, with an acid core instead of rosin. *Never* use acid-core solder for electronics work! The acid will corrode and destroy your device. By the same token, the rosin flux paste used with acid-core solder is not needed for normal electronics solder, because rosin is already in its core.

Solder comes in various diameters. A good choice for normal work is around 0.03 inches. Very small-diameter solder, in the 0.01-inch range, can be useful now and then when working with tiny parts, but not often. For most jobs, it's so undersized that you have to feed it into the work very fast to get enough on the joint, making it impractical to use. My own roll of the skinny stuff has been sitting there for a decade, and most of it is still on the roll.

Solder is like ketchup: you'll use a lot of it. Buy a 1-pound roll, because it's a much better bargain per foot than those little pocket packs of a few ounces. A pound of solder should last you a good few years.

Desoldering Tools

Removing solder to test or replace parts is as vital to repair as is soldering new ones to the board. Desoldering ranges from easy to tricky, and it's a prime opportunity for doing damage to components and the copper traces to which they're attached. Fancy desoldering stations with vacuum pumps can cost considerably more than even top-end soldering irons. For most service work, though, you don't need anything exotic. There are some low-cost desoldering options that usually do the trick.

Solder Wick

One of the best desoldering tools is *desoldering braid*, commonly called *solder wick*. It's made of very fine copper wire strands woven into a flat braid. Usually, it is coated with rosin to help solder flow into it. (You may run across some cheap wick with no rosin. Don't buy it; it doesn't work.) Wick can be purchased in short lengths on small spools from RadioShack and various mail-order companies. Electronics supply houses offer it in much longer lengths on bigger spools. As with solder, the bigger spools are the far better deal. Always be sure to keep some wick around; it's some of the most useful stuff in your workshop.

Bulbs

Another approach to solder removal is to suck it up with a rubber solder bulb, or solder sucker. Bulbs come in two forms: stand-alone and integrated with a soldering iron (see Figure 2-5). Both have their uses, but with the integrated type, you're limited by the heating power of the built-in iron, which is usually not especially strong. Stand-alone bulbs are cheap, so get one even if you also get an integrated type.

Spring-Loaded Solder Suckers

One of the handiest solder removal tools, the spring-loaded solder sucker is another inexpensive option. These bad boys have a fast, almost violent action, and are a bit harder to control than bulbs. They suck up a lot of solder in one motion, though. Get one.

Vacuum Pump Desoldering Irons

These are like integrated bulb desoldering units, except that the suction is provided by a vacuum pump instead of a bulb. The pros use these, and they're fast and powerful, but they're expensive. If you can snag a good used one at a hamfest, go for it. Just be sure the heating and vacuum systems work properly. The vacuum portions are prone to problems and worn parts, because molten solder flows through them.

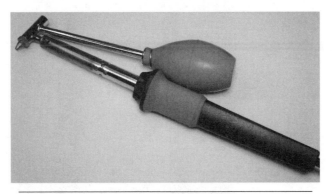

FIGURE 2-5 Bulb-type desoldering iron

Chip Quik

This is a special low-temperature solder alloy and flux kit used for desoldering surface-mount parts. When melted into the existing solder, the alloy keeps it molten at low temperatures, allowing you to get lots of pins hot enough simultaneously to remove even high-density chips with dozens of leads.

Hand Tools

The range of available hand tools seems practically infinite. Most likely, you'll build up a significant collection of them as the years go by. My own assortment fills several drawers. While nobody needs six pairs of needlenose pliers, there is a core set of tools necessary for disassembling and reassembling the items to be repaired.

Screwdrivers

Today's gadgetry uses a wide range of types and sizes of screws. Some of the screws are incredibly tiny. A set of jeweler's screwdrivers is a necessity. Both Phillips and flatblade screws are used, though Phillips types dominate. More and more, hex and Torx heads are showing up too (see Figure 2-6). The latter shapes started out as a way to prevent consumers from opening their gadgets, but the drivers have gradually become available, defeating that objective. In response, newer types have come along. One of the most recent is the Trigram, which looks like a center point with three lines radiating out toward the perimeter of the screw head. In time, those drivers will be easier to find as well.

Get a good selection of small drivers in all these form factors. At the very least, get Phillips, flat blade and Torx screwdrivers. Pick up a few medium-sized Phillips and flathead drivers too. You'll use the smaller ones much more often than the larger ones, but it pays to have as many sizes as you can find. Really big ones are rarely needed, though.

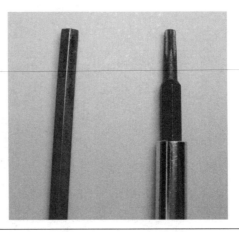

FIGURE 2-6 Hex and Torx driver tips

Cutters

Diagonal cutters, or *dikes*, are used to clip the excess lead lengths from newly installed components. Most techs also use them to strip insulation off wire. Again, smaller beats bigger. Get a couple of pairs of these things, because they tend to get bad nicks in their cutting edges and gradually become useless. A pair of dikes shouldn't cost more than about $7. Oh, and be sure the handles are insulated. They usually are.

Needlenose Pliers

A pair of needlenose pliers is essential for grabbing things, reaching into cramped spots, and holding parts steady while you solder them. A length of 2 or 3 inches from the fulcrum to the tips is about right. Any shorter and they may not reach where you need them. Any longer and they'll probably be a bit too flexible, reducing their usefulness when twisting is required. Unlike cutters, needlenose pliers rarely wear out or need replacement. Still, get two pairs so you can hold one in each hand and use them at the same time. You'll need to do that now and then.

Hemostats

They're not just for surgeons anymore! Hemostats are much like needlenose pliers, except that they lock, providing a firm grip without your having to keep squeezing the handle. Some have corrugated gripping ends, while others are smooth. Get one pair of each style. These things are indispensable for pulling a component lead from a board while heating the solder joint on the other side. They're great for installing new parts, too. You can find hemostats at most electronics and medical supply houses.

Magnifier

With the size of today's electronics, human eyes have hit their resolution limit for comfortable close-up work. It's essential that you have some magnification. Even if your spot lamp has a magnifying lens, you'll still need a head-worn magnifier, because the lamp will get in your way when it's placed between your face and a small gadget. Glass is better than plastic, which gets scratched and can even melt when situated very near a soldering iron. Be sure the magnifier you choose can be flipped up out of the way, because sometimes you need to step back a bit from the work and take in a longer view. If you wear glasses, you may need to get a magnifier with adjustable focus to keep it compatible with your eyewear.

Clip Leads

Frequently, testing involves making temporary connections. For that, nothing beats a batch of clip leads, which are wires about a foot long with alligator clips at both ends. Get at least ten, making sure that the clips are small and have rubber insulating covers.

You can buy the clips and make them yourself, but assembled ones are readily available and inexpensive. Just be aware that the premade ones are usually not soldered; the wires are merely crimped to the clips. After some use, they break inside the insulation, which leads to some head scratching when a connection doesn't produce the expected results. You can never completely trust the integrity of a clip lead. The quick test is to pull the lead taut while holding on to the clips. If broken, one end will fall apart, after which you can solder it on. In time, you'll wind up soldering all of them.

Swabs

Cotton swabs are very useful in the shop. Get the kind with paper sticks, not plastic ones. The paper type can be bent into shapes that will let you poke them into odd corners. If you can find them, also get some chamois swabs. Unlike the cotton type, chamois swabs don't leave little fibers behind. For some uses, especially cleaning video heads on VCRs and camcorders, the fibers can be problematic, and it's even possible to break a video head if the fibers get snagged on it while you clean.

Contact Cleaner Spray

There are many brands of spray, each claiming superiority, but they all do pretty much the same thing: remove oxidation and dirt from electrical contacts. One of the more popular brands is DeoxIT. RadioShack's spray is called TV-Tuner/Control Cleaner & Lubricant. Get a can or two. It's handy stuff and you'll be using it, especially if you work on older gear.

Alcohol

Alcohol can be very useful in cleaning tape paths and heads. Use isopropyl alcohol, and look for the highest percentage you can find. The 70-percent solution sold in drugstores is 30 percent water, which is bad for electronics. Some stores sell 91-percent, which is much better, and I've run across 99-percent on occasion. Don't use ethyl or any other type of alcohol. Be aware that all alcohols can damage some types of plastic rather badly. When working with alcohol, keep it away from plastic casings, LCD screens and control panels.

Naphtha

Sold in little yellow bottles as "cigarette lighter fuel" at grocery stores, and in bigger containers as "VM&P Naphtha" at hardware stores, naphtha is an amazing solvent that will effortlessly remove grime, sticker adhesive, solder rosin, tobacco tar and other general filth from just about any surface. I've never seen it harm plastic, either, not even LCD screens. It's used by dripping a very small amount on a tissue, paper napkin or swab, and then gently rubbing the surface to be cleaned. Naphtha is seriously flammable, so *never* use it on anything to which power is applied, or near flame or a hot soldering iron. It's best to use gloves, too, to keep it off your skin. Make sure of proper ventilation, because it evaporates quickly and should not be inhaled. Even if you buy a big can of it, also buy one of the little yellow bottles so you can squirt out tiny doses as needed. It takes very little naphtha to do the job, and you can always refill the bottle from the big can *outside your house* later on. Keep naphtha containers tightly closed so evaporated fumes won't build up and become hazardous in biological or fire hazard terms.

Heatsink Grease

This silicone-based grease is used between transistors, voltage regulators and other heat-producing parts and their metal heatsinks. It fills in the tiny gaps between imperfect surfaces, helping transfer heat from the part to the heatsink. Even when mica insulators are used, heatsink grease is still required for most parts. (The exception is an installation using a special rubber heat transfer gasket; most of those do not require grease.) When you replace a heatsinked part, you'll need the grease; omitting it will result in an overheated component that will quickly fail. A small tube of heatsink grease lasts a very long time, as only a thin film is required, and too much grease can actually reduce heat transfer.

Silicone grease is inappropriate for use with microprocessors and graphics chips. These hot-running parts require special silver-bearing grease, which you can find at computer stores and mail-order suppliers.

Heat-Shrink Tubing

This stuff looks like ordinary plastic tubing, but it has a wonderful trick up its synthetic rubber sleeve: it shrinks in diameter when you heat it up, forming itself around joints and damaged insulation spots in wires. It's much more permanent than electrical tape, which tends to get gooey and let go after awhile. Get some lengths of heat-shrink tubing in various small diameters.

Electrical Tape

Despite its impermanence, electrical tape still has uses in situations where tubing won't fit or can't be slipped over what needs to be insulated. Plus, for extra insulative peace of mind, you can wrap a connection in tape and then put tubing over it.

Small Cups

If you eat yogurt or pudding, start saving the little plastic cups. However you obtain them, those cups are incredibly useful for temporary storage of screws and other small parts as you disassemble machines. Make sure the cups fit into each other. Most will.

Internet Access

Not that long ago, a significant stack of reference books was required for looking up transistor types, cross-referencing replacement components and finding disassembly hints and diagrams for various products. Now we can do all that and more on the Internet! Some sites charge for schematics, but you can find some free ones. Even if you have to pay a little bit for the diagram, it may be well worth it. There's plenty of free info out there about how to take apart certain products without breaking them, too. Having a computer nearby with Net access is really handy.

Nice-to-Haves

Here are some items that can help you get repair work done more easily. You can live without 'em, but you might want to add some to your arsenal as time goes on.

Digital Camera

How's your memory? If it's imperfect, like most of ours, a digital camera can save your rear end when you look up from the bench and realize you've removed 35 screws from four layers of a laptop, and you're not sure where they all go. And what's that funny-looking piece over there? The one you took off three days ago, just before you answered the phone and took the dog for his emergency walk? Take pictures as you disassemble your devices. Use the macro lens as necessary to get clear close-up shots; a blurry photo of a circuit board does you little good. Make sure you can see which plug went in which connector, and what the shield looked like before you removed it. Experiment a bit with the flash, too, and the angles required to get decent shots without too much glare.

Power Supply

Unless it has its own AC power supply, your repair item runs either on batteries or from an AC adapter. AC adapters themselves fail often enough that you can't assume the adapter isn't the problem. So, running the device under test from a variable power supply can really help. The most important issue when choosing a supply is how much current it can provide. While a little pocket radio might eat around 50 ma (milliamps), a power amplifier or radio transmitter may require hundreds of times as much current. For most work, if you have 5 amps available, you're covered.

There are some fancy laboratory-grade power supplies with digital metering, ultra-precise regulation, and price tags to match. You don't need one. Any decent, hobby-grade supply will do.

Many of the items that demand high current are for use in the automotive environment, so a 12-volt (more typically 13.8 volts, the actual voltage of a car with its engine running) supply with 10 amps or more is great to have as well. That one doesn't need to be variable, since all auto gear runs on the same voltage.

Transistor Tester

Although it's possible to test many characteristics of transistors with a DMM, some failure modes, like excessive reverse leakage (when a small current can flow backward through a defective transistor's junctions), aren't easy to find that way. Dedicated *dynamic* transistor testers use the transistor under test as part of an oscillator, measuring how the part behaves with real signals applied to it. Under those conditions, you can measure the approximate gain, high-frequency cutoff point and leakage. Many testers can check various transistor types, including MOSFETs (metal-oxide-semiconductor field-effect transistors), junction FETs (field-effect transistors) and standard bipolars.

Basic transistor testers of this type are inexpensive and a great addition to your bench setup. They're fairly easy to make, too, and you can find diagrams in hobbyist magazines.

Capacitance Meter

Capacitors, especially the electrolytic type used in power supplies as filtering elements to smooth the output power, are some of the most trouble-prone components of all. In addition to suffering complete failures like opens and shorts, capacitors can gradually lose their ability to store energy. Worse, their internal equivalent series resistance (ESR) can rise, in which case the capacitor will measure just peachy keen on a meter, but will act like it has a big resistor between it and the circuit when it's in use. Excessive capacitor ESR is one of the most common causes of oddball circuit behaviors.

Some DMMs have built-in capacitance measurement, but ESR meters are still on the expensive side. You can get by without one, though; in Chapter 11 we're going to explore how to evaluate capacitors in operating circuits, using our good friend the oscilloscope.

Signal Generator

More useful for servicing analog equipment than digital, a signal generator lets you inject a test signal into a device's signal-processing stages to see whether doing so causes the expected effect. With today's digital devices, it's not something you'll use very often, but it has some application in simulating the clock oscillators that drive digital circuitry. Many products, like MP3 players, have both analog and digital sections, and a signal generator can come in handy with those if the audio circuitry malfunctions.

The generators are called *function generators* when they have the ability to create different kinds of waveforms, such as sine, triangle and square waves. While sine wave–only generators are usually segregated by frequency band, either audio or radio, function generators may have a wide range encompassing both, though they don't offer high-frequency ranges anywhere near those of radio-only generators. Many function generators operate from a couple of Hz (hertz, or cycles per second) to around 2 MHz (megahertz, or millions of cycles per second), while radio-frequency (RF) generators may reach hundreds of megahertz.

Frequency Counter

A frequency counter does just that: count the frequency of a signal. It does so by opening a gate for a precise period of time and counting how many cycles of the signal get through before the gate closes again. A counter is most useful when the frequency of a circuit's oscillator needs to be adjusted accurately. This is rarely the case with digital devices like cameras and computers, but it can be critical when calibrating the master oscillators that control the tuning of radio receivers and transmitters.

If you're considering getting a counter, look more at its low-frequency capabilities than at the high end, unless you plan to work on UHF or microwave systems. Many of the inexpensive counters that can hit 1 GHz (gigahertz, or billions of cycles per second) are optimized for radio work and have gate times too short to count audio frequencies accurately. (The slower the signal frequency to be counted, the longer the gate has to stay open to let enough cycles through for a proper count.) Oh, and counters are another product category, like DMMs, that may display lots of meaningless digits. Especially with the cheapies, there can be a long string of numbers to be taken with a significant grain of sodium chloride.

Frequency counters are capable of counting regular, continuous signals only; they're useless with complex, changing ones. To extract frequency information from those, you need...yup, a scope. See, I told you that darned scope was your friend!

Speaking of scopes, you'll need an extra scope probe for your counter unless you want to share one with the scope. Get one with a 10X/1X switch, as that is especially handy for counter use.

Analog Meter

The moving meter needle of an old-fashioned analog VOM offers some info to the trained eye that a modern DMM can't. Slowly fluctuating voltages, which you might encounter with, perhaps, a bad voltage regulator or a circuit pulling too much current, are even easier to see with a VOM than with an oscilloscope. Little voltage dips or spikes cause a characteristic bounce of the needle that's very informative, too. You can even get a rough estimation of an electrolytic capacitor's condition with an analog ohmmeter by watching the needle quickly rise and then slowly drop. With a DMM, such changes are just rapidly flashing numbers impossible to interpret.

A special type of VOM is known as a VTVM, for *vacuum tube volt meter*. A VTVM works like a VOM, but it contains an amplifier, making it considerably more sensitive to small signals and much less likely to steal meaningful amounts of current from the circuit under test, or *load it down*. VTVMs go back a long way, from before there were transistors, and early ones really did use vacuum tubes. Later models substituted the tubes with a very sensitive type of transistor called a *field effect transistor (FET)*. Those were known as FET-VOMs, but most people continue to call any amplified analog meter a VTVM, whether it has a tube or not. If you can find a FET-VOM, you might want to snap it up, because they're getting rare. True VTVMs are very, very old, and replacements for those small tubes are hard to find, but some working ones are still out there.

Decent VOMs, FET-VOMs and VTVMs occasionally turn up at hamfests. As long as it works, an old VOM is as good as a new one; there's not much in it to go wrong. If the meter needle moves without getting stuck and the selector switch works, you should be good to go. True VTVMs may run on AC power or on batteries, but the battery-operated units require cells nobody makes anymore, so avoid those meters. VOMs and FET-VOMs use batteries that may sit in them for years, so check inside the battery compartment to make sure an old cell hasn't leaked and corroded the contacts. Amplified meters (VTVMs and FET-VOMs) have a lot more in them than do VOMs, too, so it's best to test their functions before buying.

Many companies have made VOMs, but the best oldies were made by Simpson, which also made the best VTVMs. Some of those ancient Simpsons still go for real money, and they're worth it. You pretty much have to shoot them when you don't want them anymore. Triplett was another company that made great meters.

Isolation Transformer

It used to be that most AC-powered gear had a *linear power supply*, which shifted the incoming voltage down to a lower one through a transformer operating at the 60-Hz line frequency. The transformer, an assembly of two or more coils of wire on an iron core, had no electrical connection between its input and output; the energy was transferred magnetically. This arrangement helped with safety, because it meant that the circuitry you might touch was not directly connected to the house wiring, and thus couldn't find a path to ground through that most delicate of all resistors, you. Unfortunately, to move a lot of power required a big, heavy transformer.

Today's *switching power supplies,* or *switchers* (see Chapter 14), chop the incoming power into fast bursts to push lots of energy through a small, light transformer. They're much more dangerous to work on, because some of their circuitry is connected directly to the AC line, and it may have several hundred volts on it—and often it's the section that needs repair.

An isolation transformer is just a big, old-style AC line transformer into which you can plug your device. The transformer has a 1:1 voltage ratio, so it doesn't change the power in any way, but it isolates it from the AC line, making service of switchers a lot safer. If you're going to work on switching power supplies while they're connected to the AC line, you *must* have an isolation transformer. Many times, you can fix switchers while they're unpowered, so having an iso transformer is optional. Just don't *ever* consider working on a live switcher without one. Seriously! You don't want the power supply to wind up being the only thing in the room that's live, if ya know what I'm saying.

Stereo Microscope

With electronics getting smaller and smaller, even a head-worn magnifier may not be enough for a comfortable view. More and more, techs are using stereo (two-eye) microscopes to get a good, close look at solder pads on grain-of-salt–sized components. When you choose a microscope, get one with low magnification power. You're not trying to see bacteria on the parts! 10X to 20X should be more than enough. Anything higher and you probably won't even be able to recognize the component. A mono (one-eye) microscope can be used, but having the depth perception that comes with stereo vision can really help, especially if you're trying to solder or desolder under the microscope. To find a microscope, check eBay. Sometimes they go for surprisingly affordable prices.

Video-camera microscopes using a computer for display are becoming available, often for less than traditional optical microscopes. If you have room for a laptop on the bench, they're worth considering.

Bench Vise

Wouldn't it be wonderful to be gifted with three hands? If you have only the standard two, you may find it difficult to hold a circuit board while pulling a component lead from one side and heating the solder pad on the other. Check out the PanaVise and similar small vises designed for electronics work. Some offer attachable arms perfect for gripping the edges of circuit boards, and they let you swivel the board to whatever angle you need.

Hot-Melt Glue Gun

A small glue gun can help you repair broken cabinet parts, and a dab of hot-melt glue is also great for holding wires down. Manufacturers sometimes use it for that, and you may have to remove the glue globs to do your work. Afterward, you'll want to replace the missing globs.

Magnet on a Stick

The first time you drop a tiny screw deep into a repair project and it won't shake out, you'll be glad you bought this tool. It's useful for pulling loosened screws out of recessed holes, too. Get one that telescopes open like a rod antenna. Just keep the darned tip away from hard drives, tape heads, video head drums and anything else that could be affected by a strong magnetic field. Also, keep in mind that the metal rod could contact voltage, so never use the tool with power applied, not even if the device is turned off. Charged electrolytic capacitors can impart a jolt after power is disconnected, so keep away from their terminals too. See Chapters 3 and 7 for more about capacitors.

Cyanoacrylate Glue

Also known by the trademarks Super Glue and Krazy Glue, instant adhesive can be useful on some plastic parts. It's strong, but it has poor shear strength and is not terribly permanent, so it shouldn't be used for repair of mechanical parts that bear stress, as the repair will not last long. This type of glue is handy for holding things together while you fasten them with other, more permanent means. Just be aware that it outgases a white film as it hardens that is tough to remove, so keep it away from lenses, display screens and other surfaces where that might be a problem.

Component Cooler Spray

This stuff is colder than a witch's, um, iced tea, and it's used to put the deep freeze on suspected intermittent components, especially semiconductors (diodes, transistors and IC chips). While it might seem primitive to blast parts instead of scoping their signals, doing so can save you hours of fruitless poking around when circuits wig out only after they warm up. One good spritz will drop the component's temperature by 50 degrees or more and can reveal a thermal intermittent instantly, returning the circuit to proper operation for a few moments until it heats up again.

Data Books

Although the Internet offers lots of great service-related information, some important tidbits are still more easily found in a good old data book. Transistor cross-reference data, which you use when you need to substitute one transistor type for another because you can't get the original type, is most easily looked up in a book. So is pinout data for various ICs, voltage regulators and varieties of transistors.

Motorola, National Semiconductor and ECG used to give away reference books, but these days you'll probably have to buy them from electronics supply houses. At the least, consider getting a transistor substitution book. Of course, if it's offered as a CD-ROM or a downloadable PDF file, that's even better, assuming you have a computer near your bench.

Parts Assortment

Having a supply of commonly used components is quite handy. You can strip old boards for parts, but it's time-consuming, and you wind up with very short leads that may be difficult to solder to another board. Plus, your stash will be hit-or-miss, with big gaps in parts values. Consider buying prepackaged assortments of small resistors and capacitors or going to a hamfest and stocking up for much less money. There, you're likely to find big bags of caps, transistors, chips, resistors, and so on, for pennies on the dollar. Avoid buying transistors and chips with oddball part numbers you don't recognize, because they may be *house numbers,* which are internal numbering schemes used by equipment manufacturers. Those numbers are proprietary, and there's no way to determine what the original type number was. Thus, you can't look up the parts' characteristics, making them useless for repair work. Resistors and capacitors, luckily, almost always have standardized markings, and you can easily measure them if you have the appropriate meters.

Don't bother buying ceramic disc capacitors, because they almost never go bad, so you aren't likely to need any. If you do run into a suspicious one, you can pull its replacement off a scrap circuit board. Instead, focus on resistors, transistors, voltage regulators, fuses and, especially, electrolytic capacitors.

Electrolytic capacitors are the cylindrical ones with plastic sleeves around them and markings like "10µf 25VDC." There are many varieties of 'lytics, but some are pretty common, and substitution of similar but not identical parts is feasible in many instances. (We'll explore how to do it in Chapter 12.) Get an assortment of caps in the range of 1 to 1000 µf (microfarads), with voltage ratings of 35 volts or more. The higher the ratings, both capacitive and voltage, the larger the cap. While it's fine to replace a cap with one of a higher voltage rating, get some rated at lower voltages too, in case there isn't room on the board for the bigger part.

Stocking up on transistors is tricky because there are thousands of types. Small-signal transistors, which don't handle a lot of current, are not hard to substitute, but power transistors, used in output stages of audio amplifiers and other high-current circuits, present many challenges, and they're usually the ones that need replacement. Still, small transistors are very cheap—in the range of 5 to 25 cents each—and it's worth having some around. Get some 2N2222A or equivalent, along with some 2N3906. You may find hamfest bags of parts with similar numbers that start with *MPS* or other headers. If the number portion is *3906* or *2222*, it's pretty much the same part and will do fine.

Diodes and rectifiers are frequent repair issues, so it pays to have some on hand. The only difference between a diode and a rectifier is how much power it handles. Small-signal parts are called *diodes*, and larger ones made for use in power supply applications are dubbed *rectifiers*. Look for 1N4148 and 1N914 for the small fry, and 1N4001 through 1N4004 for the big guns.

The *bridge rectifier*, which integrates four rectifiers connected in a diamond configuration into one plastic block with four leads, is commonly used in power supplies. It's a power-handling part that fails fairly often. Get a few with current ratings in the 1- to 5-amp range and voltage ratings of 150 to 400 volts.

To house components, most of us use those metal cabinets with the little plastic drawers sold at hardware stores. Sort resistors and capacitors by value. If you have too many values for the number of compartments in the drawers, arrange the parts into ranges. For example, one compartment can hold resistors from 0 to 1 KΩ (kilohm, or thousand ohms), while the next might contain those from just above 1 KΩ to 10 KΩ. Once you learn to read the color code (see Chapter 7), plucking the part you need from its drawermates is easy.

Scrap Boards for Parts

Despite what I said about stripping old boards, you *do* want to collect carcasses for parts. No matter how large your components inventory is, the one you need is always the one you ain't got! An old VCR or radio can provide a wealth of goodies, some of which are not easily obtained at parts houses, especially at 11:30 P.M. on a Sunday, when your hours of devoted sleuthing have finally unearthed the problem—at least you think so—and you would sell parts of your anatomy for that one darned transistor, just to see if it really brings your patient back to life.

If you have room, it's easy to pile up dead gadgetry until your spouse, conscience or neighbors intervene. You're highly unlikely ever to need cabinet parts, because they won't fit anything beyond the models for which they were made, so saving entire machines is somewhat pointless and inefficient. The better approach is to remove circuit boards, knobs and anything else that looks useful, and scrap the rest. Don't bother stripping the boards; just desolder and pull off a part when you need it. If the leads are too short, solder on longer ones.

Wish List

For most service work, you can easily live without the following items, but they make for good drooling. Some advanced servicing requires them, but not often.

Inductance Meter

This meter reads the inductance value of coils, which seems like something quite useful, right? It's really rare, though, for a coil to change its inductance without failing altogether. Usually, the coil will open (cease being connected from end to end) from a melted spot in the wire, as a result of too much current overheating the windings. In the high-voltage coils used in CRT TVs and LCD backlighting circuits, insulation between coil windings can break down and arc over, causing a short between a few windings but leaving most of them intact. That will change the coil's inductance, making an inductance meter useful. To get any benefit from it however, you need to know what the correct inductance should be, and often there's no way to ascertain that unless you have a known good coil with which to compare the suspect one. That, and the fact that coils don't wear out and show gradually declining performance the way electrolytic capacitors do, accounts for the inductance meter being on the wish list, while the capacitance meter is a little higher up the chain of desire.

Logic Analyzer

An offshoot of the oscilloscope, a logic analyzer has lots of input channels but shows only whether signals are on or off. It is used to observe the timing relationships among multiple digital signal lines. Getting benefit from it requires knowledge of what those relationships should be, information rarely provided in the service manuals of consumer electronics devices. It is unlikely you'll ever need one of these.

SMT Rework Station

Surface-mount technology, known as SMT, or sometimes SMD (surface-mount device), is today's dominant style of componentry, because it makes for much smaller products and also eliminates the need for drilling hundreds of very precise holes in the circuit board. SMT stuff is somewhat hard to work on, thanks to the size scale and the lack of holes to secure a part while you solder it. The pros use SMT rework stations, which are fancy soldering/desoldering stations with custom tips that fit various kinds of chips. SMT rework stations are expensive and not hobbyist material, at least so far.

Spectrum Analyzer

This is a special type of scope. Instead of plotting voltage versus time, a spec-an plots voltage versus frequency, letting you see how a signal occupies various parts of the frequency spectrum. Used extensively in design and testing of radio transmitters, spec-ans are expensive overkill for most service work, unless RF is your thing. Ham radio operators covet these costly instruments, but you won't need one to fix normal consumer electronics devices. Besides, it is illegal to service transmitters in any way that could modify their spectral output unless you're a licensed amateur radio operator working on ham gear or you hold a radiotelephone license authorizing you to work on such things.

Chapter 3

Danger, Danger! Staying Safe

Before you start repairing electronics, get clear on one important fact: as soon as you crack open a product's case, you have left the government-regulated, "I'll sue you if this thing hurts me," coddled, protected world of consumer electronics behind. Once the cover comes off, you are on your own, and *you can get hurt or killed if you're not careful*! You've probably heard many times how dangerous CRT TV sets are to service, but don't fool yourself into thinking that today's gear is all that much safer. Even some battery-operated devices step up the voltage enough to zap the living crud out of you.

That said, you can learn to navigate all kinds of repairs safely. Let's look at a few ways you can get injured and how to avoid it, followed by the inverse: how you can damage the product you're trying to service.

Electric Shock

This is the most obvious hazard and the easiest to let happen. It might seem simple to avoid touching live connection points, but such contact happens all the time, because the insides of products are not designed for safety. Remember, you're not supposed to be in there! You may find completely bare, unprotected spots harboring dangerous voltage, and a slip of the tool can be serious.

Remove your wristwatch and jewelry before slipping your hand into a live electronic product. Yes, even a battery-operated one. Take off the wedding ring, too. They don't call metal contact points *terminals* for nothing!

In most devices, the electrical reference point called *circuit ground* is its metal chassis and/or metal shields. This is where old electrons go to die after having done their work, wending their way through the various components to get there. The trick is not to let them take you along for the ride! If you are in contact with the circuit ground point and also a point at a voltage higher than about 40 or 50 volts, you will get shocked. If your hands are moist, even lower voltages can zap you. The bodily harm from a shock arises from the current (number of electrons) passing through you, more than the voltage (their kick) itself. The higher the kick, though, the more electrons it forces through your body's resistance, which is why voltage matters. The path through your body is important as well, with the most dangerous being from hand to hand, because the current will flow across your chest and through your heart. That, of course, is one electrically regulated muscle whose rhythm you don't want to interrupt. So, it's prudent to keep your hand away from the circuit ground when taking measurements, just in case your other hand touches some significant voltage. In the old CRT TV service days, techs lived by the "one hand rule," keeping one hand behind their backs while probing for signals in a powered set. Also, don't service electronics while you're barefoot or wearing socks; you're more likely to be grounded, offering a path through your body for wayward electrons. Always wear shoes.

Switching power supplies (see Chapter 14) have part of their circuitry directly connected to the AC line. As I mentioned in Chapter 2 on the section about isolation transformers, that's a very dangerous thing, because lots of items around you in the room represent lovely ground points to which those unisolated electrons are just dying to go, and they don't mind going through you to get there. Once again, *never* work on circuitry while it is directly connected to the AC line. If there's no transformer between the AC line and the part of the circuit you wish to investigate, it's directly connected. Unplug it from the line even before connecting your scope's ground clip, because where you clip it may be at 120 volts or more, which will flow through the scope's chassis on its way to the instrument's ground connection, blowing fuses and possibly wrecking your scope.

Lots of AC-operated products have exposed power supplies, with no protection at all over the fuse and other items directly connected to the AC line. Touching one of those parts is no different than sticking a screwdriver in a wall socket. It's all too easy for the back of your hand to grant you a nasty surprise while your fingers and attention are aimed elsewhere. Even if the shock isn't serious (which it could be), you'll instinctively jerk your hand away and probably get cut on the machine's chassis. When probing in a device with an exposed supply, place something nonconductive over the board when you're not working on the supply itself. I like to use a piece of soft vinyl cut from the cover of a school notebook.

Capacitors, especially large electrolytics, can store a serious amount of energy long after power has been removed. I've seen some that were still fully charged weeks later, though many circuits will bleed their energy off within seconds or minutes. The only way to be sure a cap is discharged is to discharge it yourself. *Never* do this by directly shorting its terminals! The current can be in the hundreds of amps, generating a huge spark and sometimes even welding your tool or wire to the terminals. Worse, that fast, furious flow can induce a gigantic current spike into the device's circuitry, silently destroying transistors and chips. Instead, connect a 10-ohm resistor rated at a watt or two to a couple of clip leads, and clip them across the terminals to discharge the cap a little more slowly. Keep them connected for 20 seconds, and then remove one and measure across the cap with your DMM set to read DC voltage. It should read zero or close to it. If not, apply the resistor again until it does.

Before discharging a cap, look at its voltage rating, because the voltage on it will always be less than the rating. If a cap is rated at 16 volts, it isn't going to be dangerous. If it's rated at 150 volts, watch out. Even with the low-voltage part, you may want to discharge it before soldering or desoldering other components, to avoid causing momentary shorts that permit the cap's stored energy to flow into places it doesn't belong. Most of the time, low-voltage caps are in parts of circuits that cause the capacitors to discharge pretty rapidly once power is turned off, but not always.

The capacitance value tells you how much energy the capacitor can store. A 0.1 µf cap can't store enough to cause you harm unless it's charged to a high voltage, but when you have tens, hundreds or thousands of µf, the potential for an electrifying experience is considerable at the lower voltages you're more likely to encounter.

CRTs, especially in color TV sets, act like capacitors and have low enough leakage to store the high voltage applied to their anodes (the hole in the side with the rubber cap and the thick wire coming from it) for months. There isn't much capacitance, thus not a lot of current, but the voltage is so high (anywhere from 25,000 to 50,000 volts!) that what there is will go through you fast and hard enough to cause a large, sad family gathering about a week later. CRTs are going the way of the dodo, so you probably won't work with them anyway. Just be very, very careful if you do. The terminals at the back of the tube carry some pretty high voltages too.

The backlighting circuits of LCD monitors and TVs, along with much of the circuitry of plasma TVs, operate at high enough voltages to be treated with respect. It's unwise to try to measure the output of a running backlighting circuit at the point where it connects to the fluorescent lamp tube unless you have a high-voltage probe made for that kind of work. Without one, you may get shocked from the voltage exceeding the breakdown rating of your probe, you're likely to damage your DMM or scope, and the added load also may blow the backlighting circuit's output transistors.

Speaking of lamps, the high-pressure mercury vapor arc lamps used in video projectors are "struck," or started, by putting around a kilovolt on them until they arc over, after which the voltage is reduced to about 100 volts. Keep clear of their connections during the striking period, and don't try to measure that start-up voltage.

Physical Injury

The outsides of products are carefully designed to be user-safe. Not so the insides! It's easy to get sliced by component leads sticking up from solder joints, by the edges of metal shields, and even by plastic parts. Move deliberately and carefully; quickly shoving a hand into nooks and crannies leads to cuts, bleeding and cursing. That said, it still happens often enough that my years of tech work led me to coin the phrase, "No job is complete without a minor injury."

CD and DVD players and recorders (especially recorders) put out enough laser energy to harm your eyes, should you look into the beam. Video projectors use lamps so bright that you *will* seriously damage your vision by looking directly at them. The lamps and their housings get more than hot enough to burn you, and hot projection lamps are very fragile, so don't bounce the unit or hit anything against it while it's operating. An exploding lamp goes off like a little firecracker, oh-so-expensively showering you with fine glass particles and a little mercury, just for extra effect.

Speaking of eyes, yours will often be at rather close range to the work. Much of the time, you'll be wearing magnifying lenses offering some protection from flying bits of wire or splattered molten solder. When the magnifiers aren't in use, it's a good idea to wear goggles, especially if you don't wear glasses. Excess component leads clipped with diagonal cutters have an odd, almost magnetic tendency to head straight for your corneas at high speed. Solder smoke also likes to visit the area, and it can be pretty irritating.

You can hurt your ears, too, particularly when working on audio amplifiers with speakers connected. Touching the wrong spot may produce a burst of hum or a squeal loud enough to do damage when your ears are close to the speakers. That sort of thing happens mostly with musical instrument amplifiers, because their speakers are right in your face when you work on them, and those amps pack quite a wallop. Even a 15-watt guitar amp can get painfully loud up close. Don't think turning the volume knob down will protect you; there are plenty of places you can touch that will produce full power output regardless of the volume control's position.

Other opportunities for hearing damage involve using headphones to test malfunctioning audio gear. Even a little MP3 player with just a few milliwatts (thousandths of a watt) of output power can pump punishingly loud noises into your ears, particularly when ear buds that fit into the ear canals are used. If you must wear headphones to test a device, always use the over-the-ear type, and pull them back so they rest on the backs of your earlobes. That way, you can hear what's going on, but unexpected loud noises won't blast your eardrums.

Breathing in solder smoke, contact cleaner spray and other service chemicals isn't the healthiest activity. Keep your face away when spraying. When you must get close while soldering, holding your breath before the smoke rises can help you avoid inhalation.

Your Turn

Sure, electronics can hurt you, but you can hurt the equipment too. Today's devices are generally more delicate than those of past decades. It was pretty hard to damage a vacuum tube circuit with anything short of a dropped wrench hitting the glass. Today's ultra-miniaturized circuitry is an entirely different slab of silicon. Here are some ways you can make a mess of your intended repair.

Electrical Damage

Working with powered circuits is essential in many repair jobs. You can't scope signals when they're off! Poking around in devices with power applied, though, presents great opportunity to cause a short, sending voltages to the wrong places and blowing semiconductors, many of which cannot withstand out-of-range voltages or currents for more than a fraction of a second. One of the easiest ways to trash a circuit is to press a probe against a solder pad on the board, only to have it slip off the curved surface when you look up at your test instrument, and wind up touching two pads at the same time.

Any time you stick a probe on a solder pad, be very aware of this potential slip. At some point, it'll happen anyway, I promise you. Luckily, many times it causes no harm. Alas, sometimes the results are disastrous. If you experience this oops-atronic event and the circuit's behavior suddenly changes, and toggling the power doesn't restore it to its previous state, assume you did some damage.

Another common probing problem occurs when a scope probe is too large for where you're trying to poke it, and its ground ring, which is only a few millimeters from the tip, touches a pad on the board, shorting it to ground. Again, sometimes you get away with it, sometimes you don't. In small-signal circuit stages, it's more likely to be harmless. In a power supply, well, you don't want to do it, okay? It can be helpful in tight circumstances to cut a small square of electrical tape and poke the end of the probe through it, thus insulating the ground ring.

When operating a unit with your bench power supply, there are several things you can get wrong that will wreck the product. First and foremost, don't connect positive and negative backward! Nothing pops IC chips faster than reversed polarity. Products subjected to it are often damaged beyond repair.

Some devices, especially those intended for automotive use, have reverse-polarity protection diodes across their DC power inputs. The diode, deliberately connected backward with its anode to – and its cathode to +, doesn't conduct as long as the power is correctly applied. When polarity is reversed, the diode conducts, effectively shorting out the power input and usually blowing the power supply's fuse, protecting the product's sensitive transistors and chips from backward current. If the power supply has a lot of current available, the diode may rapidly overheat and short, requiring its replacement, but the rest of the unit should remain unharmed.

Few battery-operated products have protection diodes. Reverse-polarity protection is usually accomplished mechanically in the battery compartment by a recessed terminal design that prevents the battery's flat negative terminal from touching the positive contact. The AC adapter jack probably isn't protected either, because it's assumed you will use the adapter that came with the product.

So, be very careful to connect your power supply the right way around, and never hook it up while power is turned on, lest you even momentarily touch the terminals with your clips reversed.

Be sure you've set your power supply's voltage correctly, too. Undervoltage rarely causes damage, but overvoltage is likely to do so if it's applied for more than a few seconds. We're not talking millivolts here; if you're within half a volt or so, that's usually good enough. A decade ago, most items ran on unregulated linear-type adapters and did the regulating internally, so they were fairly tolerant of having excessive voltage coming in, and you were fine if you were within 2 or 3 volts. These days, more and more products are using regulated, switching-type AC adapters with very steady voltage outputs, so the gadgets expect a pretty accurate voltage.

It's possible to cause electrical damage when taking measurements, even if you don't slip with the probe. While scopes and meters have *high impedance* inputs that won't present any significant load to most circuits, sometimes you may be tempted to connect resistors or capacitors across points to gauge the effects. That can be an effective diagnostic technique in some cases, but it should be used with caution because you can pull too much current through some other component and blow it.

Other times, you may want to connect a voltage to a point to see if it restores operation. That, too, can be useful, but it requires consideration of the correct voltage and polarity, the amount of current required, and exactly where that energy will go. Get one of those things wrong, and you could let some of that magic smoke out of the unit's components, with predictable consequences.

When your scope is set to AC coupling (see Chapter 6), it inserts a capacitor between the probe and the rest of the scope. After you probe a point with a DC voltage on it, that voltage remains on the capacitor and will be discharged back through the probe into the next point you touch. The amount of current is very, very small, but if you touch a connection to an especially sensitive IC chip or transistor that can't handle the stored voltage, you could destroy the part in the short time it takes to discharge the scope's capacitor. When using AC coupling, touch the probe to circuit ground between measurements to discharge the cap and prevent damage to delicate components.

A static charge from rubbing your shoes across the carpet, or just the dry air of winter, can put hundreds of volts, or even a few thousand, on your fingertips and any tools you're holding. If you think you could be charged, and especially any time you handle CMOS chips, MOSFET transistors, memory cards or other sensitive semiconductors, touch a grounded object first. The metal case of your bench power supply or analog scope should do the trick, as long as it's plugged into a three-wire outlet. I don't recommend using a digital scope as a discharge point because it has plenty of sensitive chips inside, and you sure don't want to damage those!

Physical Damage

There are lots of ways you can break things when you're inside a machine. One of the easiest is to tear a ribbon cable or snap off a critical part while disassembling the device. Some products pop apart easily but may have hidden risks. I once serviced a video projector cleverly designed to snap open without a single screw, but a tiny ribbon cable linked the top and bottom, and I was lucky that it popped out of its connector without being torn in half when I removed the top case. Had I pulled a little harder, I'd probably have done damage difficult to repair. As careful as I was trying to be, I still didn't see that darned ribbon until it was too late.

Small connectors of the sorts used on laptop motherboards and pocket camcorders can be torn from the circuit board. Today's products are soldered by machine, and the soldering to connectors isn't always the greatest, because they are a bit larger than the components, so they don't get quite as warm during soldering. A little too much pressure when you disconnect the cable, and the connector can come right off the board. Depending on the size scale of its contacts, it may be impossible to resolder it. Most ribbon connectors have a release latch you must flip up or pull out before removing the ribbon. Always look for it before pulling on the cable. See Figures 3-1 and 3-2.

Your soldering iron, that magic instrument of thermo-healing, can also do a lot of damage, especially to plastic. The sides of the heating element can easily press against plastic cabinetry when you have to solder in tight places, melting it and ruining the unit's cosmetics.

FIGURE 3-1   Ribbon connector with flip-up latch

FIGURE 3-2   Ribbon connector with pull-out latch

Finally, be careful where you press your fingers. Most circuitry is fairly hardy, but some components, including video heads, meters, speaker cones, microphone diaphragms, DLP projector color wheels and CD/DVD laser optical heads, just can't stand much stress and will break if you push on them even with moderate force.

You Fixed It! Is It Safe?

After repair, it's your duty as a diligent device doctor to ensure the product is safe to use. One common error resulting in an unsafe repair job is neglecting to put everything back the way it was. If you have internal shields and covers or other items left over after you close the unit, you'll need to open it back up and put them where they belong. Manufacturers don't waste a single penny on unnecessary parts, so you know they're important!

It's easy to touch wiring and melt insulation with the side of your soldering iron while you're concentrating on soldering components. In units with lots of wires, it can happen, despite your best intention to be careful. You might not notice doing it if you're focusing on the action at the tip, but the smoke and smell will alert you. Should you do this, fix the damage immediately; don't wait until the repair is over. For one thing, you may have created a short or a lack of insulation that could cause damage or injury when power is applied. For another, you might forget later and close the unit up in that condition.

Patching melted insulation can be as easy as remelting it to cover the wire, in the case of low-voltage, signal-carrying wires with only small melted spots. Or it might require cutting, splicing and heat-shrink tubing if the wire handles serious voltage, or if the damage is too great. Remember, electrical tape will come off after awhile, so never depend on it for long-term safety.

If you've replaced power-handling components like output transistors or voltage regulators, be sure to test the unit for proper operation and excess heat. Older stereo amps and receivers, for instance, sometimes require bias adjustments when the output transistors are changed. If you don't set the bias correctly, the unit will work for awhile, but it may overheat badly. Let it run on the bench for a few hours at normal listening volume and see how hot it gets. Be sure what you've fixed is really working properly before you close it up.

When the product has an AC cord, take a look at it and run your hand along its entire length, checking for cuts. Naturally, you want to live, so unplug it before doing this! You'll be amazed at how many frayed, cut and pet-chewed cords are out there. Replace the cord or repair it as seems appropriate for its condition, paying extra attention to a good, clean job with proper insulation. If the damage is only to one wire, it's easier to fix than if both sides are involved, because at least the two wires can't short to each other. With a damaged AC cord, I like to use both electrical tape and heat-shrink tubing over it.

Chapter 4

I Fix, Therefore I Am: The Philosophy of Troubleshooting

Imagine if your doctor saw you as a collection of organs, nerves and bones, never considering the synergistic result of their working together, supplying each other with the chemicals and signals necessary for life. No organ could survive on its own, but together they make a living, breathing, occasionally snoring you! Now consider how tough it'd be to solve a murder case without considering the motives, personalities and circumstances of the victim and all potential suspects. The knife is right there next to the body, but anybody on earth could have done the crime. Why was the victim killed? Who knew him? Who might have wanted him dead?

Troubleshooting, which involves skills somewhat like those of doctors and detectives, is a lot like that. You can think of an electronic device as a bunch of transistors, chips and capacitors stuffed into a box, and sometimes that's enough to find simple failures. Taking such a myopic view, though, limits you to being a mediocre technician, one who will be stumped when the problem isn't obvious. To be a top-notch tech requires consideration of the bigger picture. Who made this product, and what were the design goals? How is it supposed to work? How do various sections interact, and what is the likely result of a failure of one area on another?

Machines are *systems*. Being built by humans, they naturally reflect our biological origins, with cameras for eyes, microphones for ears, speakers for larynxes and microprocessors for brains. Even the names of many parts sound like us: tape recorders, hard drives, and optical disc players have heads, turntables have arms, chips have legs and picture tubes have necks. Some products even exhibit personalities, or at least it feels that way to us. Their features and quirks can be irritating, humorous or soothing. Their failures are much like our own, too, with symptoms that may be far removed from what's causing them, thanks to some obscure interaction that nobody, not even the circuit's designer, could have foreseen.

The more you come to understand how devices work at the macro level, the more sense their problems will make. The more you can consider products as metal and silicon expressions of human thinking, the better sleuthing skills you will attain. Before we get to the nitty-gritty of transistors, current flow and signals, let's put on our philosophers' hats and become the Socrates of circuitry, the Erasmus of electrons. Let's look at why products work and why they don't, and how to avoid some of the common pitfalls developing techs encounter. Let's become one with the machines.

Why Things Work in the First Place

When you get a few thousand parts together and apply power to them, they can interact in many ways. The designing engineer had one particular way in mind, but that doesn't mean the confounded conglomeration of components will cooperate!

Analog circuitry has a wider range of variation in its behaviors than does digital, but even today's all-digital gear can be surprisingly inconsistent. I've witnessed two identical laptop computers running exactly the same software, with exactly the same settings, but drawing significantly different amounts of current from their power supplies. I've also seen all kinds of minor variations in color quality between identical digital still and video cameras. I remember a ham radio transceiver whose digital control system exhibited a bizarre, obscure behavior in its memory storage operation that no other radio of that model was reported to have, and I never found any bad parts that might explain the symptom. I finally had to modify the radio to get it to work like all the others.

Sure, you string a few gates together and you will be able to predict their every state. Get a few thousand or more going, run them millions of times per second, and mysterious behaviors may start to crop up.

It's useful to think of all circuitry as a collection of resistors impeding the passage of current from the power supply terminal to circuit ground. As the current trickles through them, it is used to do work, be it switching the gates in a microprocessor, generating laser light for a disc player, or spinning the disc. Electrons, though, are little devils that will go anywhere they can. If there's a path, they'll find it. Malfunctions can be considered either as paths that shouldn't be there or a lack of paths that should.

In essence, when machines work properly, it's because they *have no choice*. The designer has carefully considered all the possible paths and correctly engineered the circuit to keep those pesky electrons moving along only where and when they should, locking out all possible behaviors except the desired one. When choice arises, through failing components, user-inflicted damage or design errors, the electrons go on a spree like college students at spring break, and the unit lands on your workbench.

Products as Art

A machine is an extension of its designer much as a concerto is an extension of its composer. Beethoven sounds like Beethoven, and never like Rachmaninoff, because Ludwig's bag of tricks and way of thinking were uniquely his, right? It's much the same with products. In this case, however, they tend to have unifying characteristics more reflective of their manufacturing companies than of a specific person. Still,

I suspect that an individual engineer's or manager's viewpoints and preferences set the standard, good or bad, which lives on in a company's product line long after that employee's retirement.

Understanding that companies have divergent design philosophies and quirks may help your repair work, because you can keep an eye on issues that tend to crop up in different manufacturers' machines. You may notice that digital cameras from one maker have a high rate of imaging chip failures, so you'll go looking for that instead of some other related problem when a troublesome case hits your bench. Or perhaps you've found that tape-type camcorders from a particular company often have mechanical loading problems because that manufacturer uses loading arms and other metal structures in the tape transport that are too thin, so they bend.

When you've fixed enough products, you'll begin to recognize what company made a machine just by looking at its circuit board or mechanical sections. The layouts, the styles of capacitors, the connectors, and even the overall look of the copper traces on a board are different and consistent enough to be dead giveaways.

If It Only Had a Brain

Continuing our anatomical analogy, yesterday's tech product was like a zombie. Perhaps it had an ear (microphone), some memory (recording tape) and a mouth (speaker). Each system did its simple job, with support from a stomach (power supply) and some muscles (motors, amplifiers).

What was missing was a brain. Today's gear is cranium-heavy, laden with computing power. Gone are simple mechanical linkages to control sequencing and movement of mechanisms. Instead, individual actuators move parts in a sequence determined by software, positional information gets fed back to the microprocessor, and malfunctions might originate in the mechanics, the sensors, the software, or some subtle interaction of those elements. No longer are there potentiometers (variable resistors) to set volume or brightness; buttons signal the brain to change the parameters. Heck, most gadgets today don't even have "hard" on/off switches that actually disconnect power from the circuitry. Instead, the power button does nothing more than send a signal to the microprocessor, requesting it to energize or shut down the product's circuitry.

In addition to the brain, many modern products have nervous systems consisting of intermediary chips and transistors to decode the micro's commands and fan them out to the various muscles, and organs doing the actual work. Failures in these areas can be tough to trace, because their incoming signals from the computer chip are dependent on tricky timing relationships between various signal lines. This is a profound shift from the old way of building devices, and it adds new layers of complication to repair work. Is the circuit not working due to its own malfunction, or is it playing dead because the micro didn't wake it up?

Today's machines are complete electrono-beings with pretty complex heads on their shoulders. Some offer updatable software, while many have the coding hardwired into their chips. Which would you like to be today: surgeon or psychiatrist?

The Good, the Bad and the Sloppy

It's easy for an experienced tech to tell when a repair attempt has been made by an unqualified person. The screws will be stripped, or there will be poorly soldered joints with splashes of dripped solder lying across pads on the board. Wires may be spliced with no solder and, perhaps, covered in cellophane tape, if at all. Adjustments will be turned, insulation melted, and so on. In a word: sloppiness.

That might sound exaggerated, but I used to run into it a lot when I worked in repair facilities. Most shops have policies of refusing to work on items mangled by amateurs, so discovery of obvious, inept tampering was followed by a phone call to the item's owner, who would stubbornly insist that the unit had never been apart and had simply quit working. Um, right, Sony used Scotch tape to join unsoldered wires. Sure, buddy. I remember one incident in which I refused to repair a badly damaged and obviously tampered-with shortwave radio. The owner was so angry that he called my boss and tried to have me fired! The boss took one look inside the set, clapped me on the back, laughed, and told the guy to come pick up his ruined radio and go away. Don'tcha wish all bosses were that great? The key to performing a proper, professional-quality repair job is meticulous attention to detail. Think of yourself as a surgeon, for that's exactly what you are. You are about to open up the body of this mechanical "organism" and attempt to right its ills. As the medical saying goes, "First, do no harm." Now and then, repair jobs go awry and machines get ruined—it happens even to the best techs, though rarely—but your aim is to get in and back out as cleanly as possible. In Chapters 9 through 13, we'll explore the steps and techniques required for proper disassembly, repair and reassembly.

Mistakes Beginners Make

Beyond sloppy work, beginners tend to make a few conceptual errors, leading to lots of lost time, internal damage to products, and failure to find and fix the problem. Here are some common quagmires to avoid.

Adjusting to Cover the Real Trouble

Analog devices often have adjustments to keep their circuit stages producing signals with the characteristics required for the other stages to do their jobs properly. TVs and radios are full of trimpots (variable resistors), trimcaps (variable capacitors) and tunable coils, and their interactions can be quite complex. With today's overwhelmingly digital circuits, adjustments are much less common. Many are performed in software with special programming devices to which you won't have access, but some good-old-fashioned screwdriver-adjustable parts still exist. Power supplies usually have voltage adjustments, for instance, and earlier-generation CD players were loaded with servo adjustments to keep the laser beam properly focused and centered on the track. Even a digital media receiver may have tunable stages in its radio sections.

It can be very tempting to twiddle with adjustments in the hope that the device will return to normal operation. While it's true that circuits do go out of alignment—if they didn't, the controls wouldn't be there in the first place—that is a gradual process. It *never* causes drastic changes in performance. If the unit suddenly won't do something it did fine the day before, it's not out of adjustment, it's *broken*. Messing with the adjustments will only get you into trouble later on when you find the real problem, and now the machine really is way out of alignment, because you made it that way. Leave those internal controls alone! Turn them only when you're certain everything else is working, and then only if you know precisely what they do and have a sure way to put them back the way they were, just in case you're wrong. Marking the positions of trimpots and trimcaps with a felt-tip marker before you turn them can help, but it's no guarantee you will be able to reset a control exactly to its original position. There's too much mechanical play in them for that technique to be reliable. In some cases, close is good enough. In others, slight misadjustments can seriously degrade circuit performance.

I once worked on a pair of infrared cordless headphones with a weak, distorted right channel. After some testing, it was clear that the transmitter was the culprit, and its oscillator for that channel had drifted off frequency. A quick adjustment and, sure enough, the headphones worked fine for a little while. Then the symptom returned. The real problem: a voltage regulator that was drifting with temperature. Luckily, readjusting the oscillator was easy after the new part was installed. When multiple adjustments have been made, it can be exceedingly difficult to get them back in proper balance with each other.

Making the Data Fit the Theory

Most techs have been guilty of this at some time. In my early years, mea culpa, that's for sure. You look at the symptoms, and they seem to point to a clear diagnosis—all except for one. You fixate on those that make sense, convince yourself that they add up, and do your best to ignore that anomaly, hoping it's not significant. Trust me, it is, and you are about to embark on a long, frustrating hunting expedition leading to a dreary dead end. Always keep this in mind: *If a puzzle won't fit together, there's a piece missing!* There's something you don't know, and that is what you should be chasing. Often, the anomaly you're pushing aside is the real clue, and overlooking it is the worst mistake you can make. Many maddening hours later, when you finally do solve the mystery, you'll think to yourself, "Why didn't I consider how that odd symptom might be the key to the whole thing? It was right in front of me from the start!" Ah, hindsight.... Nobody needs glasses for that.

Going Around in Circles

Sometimes you think you've found the problem, but trying to solve it creates new problems, so you go after those. Those lead to still more odd circuit behavior, so off you go, around and around until you're right back where you started. When addressing symptoms creates more symptoms, take it as a strong hint that you are on the wrong track. It's incredibly rare for multiple, unrelated breakdowns to occur. Almost always, there is one root cause of all the strangeness, and it'll make total sense once you find it. "Oh, the power supply voltage was too low, and that's why the focus wouldn't lock and the sled motor wouldn't make the laser head go looking for the track." If you're lucky, you'll have discovered that before you've spent hours fiddling with the limit switches and the control circuitry, tracing signals back to the microprocessor. Again, if the puzzle won't fit together, find that missing piece!

That's How It Goes

As with illness in the human body, just about anything can go wrong with an electronic device. Problems range from the obvious to the obscure; I've fixed machines in 5 minutes, and I've run across some oddball cases for which a diagnosis of demonic possession seemed appropriate! These digital days, circuitry is much more reliable than in the old analog age, yet modern gear often has a much shorter life span. How can both of those statements be true?

Today's products are of tremendously greater complexity, with lots of components, interconnections and interactions, so there's more to go wrong. Unlike the hand-soldered boards filled with a wide variety of component types we used to have, today's small-signal boards, with their rows of surface-mounted, machine-soldered chips, don't fail that often. But with so much more going on, they include complicated power supplies and a multitude of connectors and ribbon cables. Plus, some parts work much harder than they used to and wear out or fail catastrophically from the stress. And thanks to the rapid pace of technological change, the competition to produce products at bare-bones prices and the high cost of repair versus replacement, extended longevity is not the design goal it once was. Manufacturers figure you'll want to buy a new, more advanced gadget in a couple of years anyway. Contrary to popular myth, nobody deliberately builds things to break. They don't have to; keeping affordable products working for long periods is tough enough. Keeping expensive items functioning isn't easy either! Laptop computers, some of the costliest gadgets around, are also some of the most failure-prone, because they're very complex and densely packed, and they produce plenty of heat.

It may seem like electronic breakdowns are pretty random. Some part blows for reasons no one can fathom, and the unit just quits. That does happen, but it's not common. Oh, sure, when you make millions of chips, capacitors and transistors, a small number of flawed ones will slip through quality control, no matter how much testing you do. It's a tiny percentage, though. Much more often, products fail in a somewhat predictable pattern, with a cascading series of events stemming from well-recognized weaknesses inherent in certain types of components and construction techniques. In other words, nothing is perfect! Let's look at the factors behind most product failures.

Infant Mortality

This rather unpleasant term refers to that percentage of units destined to stop working very soon after being put into service. Imperfect solder joints, molecular-level flaws in semiconductors and design errors cause most of these. While many products are tested after construction, cost and time constraints prohibit extensive "burning in" of all but very expensive machines. Typical infant mortality cases crop up within a week or two of purchase and land in a warranty repair center after being returned for exchange. So, you may never see one unless you bought something from halfway around the world, and it's not worth the expense and trouble to return it. Or perhaps the seller refuses to accept it back, and you get stuck with a brand-new, dead device you want to resurrect.

Mechanical Wear

By far, moving parts break down more often than do electronic components. Hard drives, VCR and camcorder mechanisms, disc trays, laser head sleds and disc-spinning motors are all huge sources of trouble.

Bearings wear out, lubrication dries up, rubber belts stretch, leaf switches (internal position-sensing switches) bend, nylon gears split, pet hairs bind motor shafts, and good old wear and tear grind down just about anything that rubs or presses against anything else. If a device has moving parts and it turns on but doesn't work properly, look at those first before assuming the electronics behind them are faulty. For every transistor you will change, you'll fix five mechanical problems.

Connections

Connections are also mechanical, and they go bad very, very often. Suspect any connection in which contacts are pressed against each other without being soldered. That category includes switches, relays, plugs, sockets, and ribbon cables and connectors.

The primary culprit is corrosion of the contacts, caused by age and sometimes, in the case of switches and relays, sparking when the contacts are opened and closed. Also, a type of lubricating grease used by some manufacturers on leaf switches tends to dry out over time and become an effective insulator. If the contact points on a leaf switch are black, it's a good bet they are coated with this stuff and are not passing any current when the switch closes. See Figure 4-1.

FIGURE 4-1   Leaf switch

A particularly nasty type of bad connection occurs in multilayer printed circuit boards. At one time, a dual-layer board, with traces on both sides, was an exotic construct employed only in the highest-end products. Today, dual-layer boards are pretty much standard in larger, simpler devices, while smaller, more complex gadgets may utilize as many as six layers!

The problems crop up in the connections *between* layers. Those connections are constructed differently by the various manufacturers. The best, most reliable style is with plated-through holes, in which copper plating joins the layers. As boards have shrunk, plated-through construction has gotten more difficult, resulting in a newer technique that is, alas, far less reliable: holes filled with conductive glue. This type of interconnect is recognizable by a raised bump at the connection point that looks like, well, a blob of glue (see the translucent glue over the holes in Figure 4-2). Conductive glue can fail from flexure of the board, excessive current and repeated temperature swings. Repairing bad glue interconnects is hard, too. I always cringe when I see those little blobs.

Solder Joints

Though they're supposed to be molecularly bonded and should last indefinitely, solder joints frequently fail and develop resistance, impeding or stopping the current. When it happens in small-signal, cool-running circuitry, it's usually the fault of a flaw in the manufacturing process, even if it takes years to show up. Heat-generating components like output transistors, voltage regulators and video processing chips on computer motherboards can run hot enough to degrade their solder joints gradually without getting up to a temperature high enough to actually melt them. Over time, the damage gets done and the joints become resistive or intermittent.

FIGURE 4-2 Conductive glue interconnects

Many bad solder joints are visually identifiable by their dull, mottled or cracked appearance. Now and then, though, you'll find one that looks perfect but still doesn't work, because the incomplete molecular bonding lies beneath the surface. Bonding may be poor due to corrosion on the lead or pad of the soldered components; solder just won't flow into corroded or oxidized metal. When you go to resolder it, you'll have problems getting a good joint unless you scrape things clean first, after removing the old solder.

Heat Stress

Heat is the enemy of electronics. It's not an issue with most pocket-sized gadgets, but larger items like video projectors, TVs and audio amplifiers often fail from excessive heating. So do backlight inverters (the circuits that light the fluorescent lamps behind LCD screens) and computer motherboards. Power supplies create a fair amount of heat and are especially prone to dying from it.

Overheating from excessive current due to a shorted component can quickly destroy semiconductors and resistors, but normal heat generated by using a properly functioning product can also gradually degrade electrolytic capacitors, those big ones used as power supply filtering elements, until they lose most of their capacitance.

Electrical Stress

Running a device on too high a voltage can damage it in many ways. The unit's voltage regulator may overheat from dissipating all the extra power, especially if it's a linear regulator. Electrolytic capacitors can short out from being run too close to, or over, their voltage limits. Semiconductors with inherent voltage requirements may die very quickly.

Overvoltage can be applied by using the wrong AC adapter, a malfunctioning adapter, a bad voltage regulator, or using alkaline batteries in a device made for operation only with nickel-metal hydride (NiMH) rechargeable cells. Those cells produce 1.2 volts each, compared to the 1.5 volts of alkalines. So, with four cells, you get 6 volts with the alks, compared to the 5 volts the device expects. Most circuits can handle that, but some can't.

Believe it or not, a few products can be damaged by too *little* voltage. Devices with switching power supplies or regulators compensate for the lower voltage by pushing more current through their transformers with wider pulses, to keep the output voltage at its required level. That can cause overheating of the rectifiers and other parts converting the pulses back to regulated DC.

The ultimate electrical stress is a lightning strike. A direct strike, as may occur to a TV or radio with an outdoor antenna that gets zapped, or from a hit to the AC line, will probably result in complete destruction of the product. Now and then, only one section is destroyed and the rest survives, but don't bet on it. Lightning cases tend to be write-offs; you don't even want their remains in your stack of old boards, lest their surviving parts have internal damage limiting their life spans.

Power surges, in which the AC line's voltage rises to high levels only momentarily, can do plenty of damage. Such surges are sometimes the result of utility company errors, but more often lightning has struck nearby and induced the surge without actually hitting the line, or it has hit the line far away. Often, the power supply section of the product is badly damaged but the rest of the unit is unharmed.

When too much current passes through components, they overheat and can burn out, sometimes literally. Resistors get reduced to little shards of carbon, and transistors can exhibit cracks in their plastic cases. The innards, of course, are wiped out. This kind of stress rarely occurs from outside, because you can't force current through a circuit; that takes voltage. When overcurrent occurs, it's because some other component is shorting to ground, pulling excessive current through whatever is connected in series with it.

Nothing kills solid-state circuitry quite as fast as reversed polarity. Many semiconductors, and especially IC chips, can't handle current going the wrong way for more than a fraction of a second.

Batteries can be installed backward. Back when 9-volt batteries were the power source of choice for pocket gadgets, all it took was to touch the battery to the clip with the male and female contacts the wrong way around and the power switch turned on. Now that AAA cells and proprietary rechargeable batteries run our diminutive delights, that kind of error occurs less often, because it's routine for designers to shape battery compartments to prevent reversed contacts from touching, but it still happens on occasion.

By far, the most frequent cause of reversed polarity is an attempt to power a device from the wrong AC adapter. Today, most AC adapters connect positive to the center of their coaxial DC power plugs and negative to the outside, so that an automotive cigarette lighter adapter made for the same gadget doesn't present the risk of having positive come in contact with the metal car body, which would cause a short and blow the car's fuse. At one time, though, many adapters had negative on the center instead, and a few still do on items like answering machines, which will never be used in cars. Even from the same manufacturer, both schemes may be employed on their various products.

The train wreck occurs when the user plugs in the wrong adapter, and it happens to have the plug wired opposite to what the device wants. Damage may be limited only to a few parts in the power supply section, or it can be extreme, taking out critical components like microprocessors and display drivers.

Not all electrical stress is caused by external factors or random component failures. Sometimes design errors are inherent in a product, and their resulting malfunctions don't start showing up until many units are in the field for awhile. When a manufacturer begins getting lots of warranty repair claims for the same failure, the alarm bells go off, and a respectable company issues an ECO, or *engineering change order,* to amend the design. Units brought in for repair get updated parts, correcting the problem. A really diligent manufacturer will extend free ECO repairs beyond the warranty period if it's clear that the design fault is bad enough to render all or most of the machines in the field inoperative, or if any danger to the user could be involved.

At least that's how it's supposed to work. Sometimes companies don't want to spend the money to fix their mistakes, so they simply deny the problem. Or, if only some machines exhibit the symptom, they're treated as random failures, even though they're not. Perhaps it takes a certain kind of use or sequence of operations for the issue to become evident, and the manufacturer genuinely believes the design is sound. And some units aren't used often enough to have experienced the failure, though they will eventually, masking its ultimate ubiquity.

Any of these situations can result in your working on a product with a problem that will recur, perhaps months later, after you've properly solved it. If the thing keeps coming back with the same issue, suspect a defective design.

Physical Stress

Chips, transistors, resistors and capacitors can take the physical shock of being dropped, at least most of the time. Many other parts can't, though. Circuit boards can crack, especially near the edges and around screw holes and other support points. Larger parts, with their greater mass, can break the board areas around them. That happens often with transformers and big capacitors. On a single or dual-layer board, you may be able to bridge foil traces over the crack with small pieces of wire and a little solder if the traces are not too small. With a multilayer board, you may as well toss the machine on the parts pile, because it's toast.

LCDs, fluorescent tubes and other glass displays rarely survive a drop to a hard surface. The very thin, long fluorescent lamps inside laptop screens are particularly vulnerable to breakage. If you run across a laptop with no backlight, don't be too surprised if it got dropped and the lamps are broken inside the screen. I've seen that happen with no damage to the LCD itself being evident.

If you leave carbon-zinc or alkaline batteries installed long enough, they will leak. Not maybe, not sometimes, they *will*. Devices like digital cameras take a fair amount of current and get their batteries changed often, but those with low current demand, such as digital clocks and some kids' toys, may have the same batteries left in them for years. Remote controls are prime candidates for battery leakage damage, because most people install the cheap, low-quality batteries that come with them and never change them; their very low usage ensures those junky cells will be in there until they rot.

Once the goo comes out, you're in for a lot of work cleaning up the mess. They don't call them *alkaline* batteries for nothing! The electrolyte is quite corrosive and will eat the unit's battery springs and contacts. If the stuff gets inside and onto the circuit board, that's where the bigger calamity goes down. Copper traces will be eaten through, solder pads corroded, and those pesky circuit board layer interconnects will stop working. No shop will try to repair such damage, but you might want to give it a go if the device is expensive or hard to replace.

People sit on their phones, PDAs and digital cameras fairly often, resulting in cracked LCDs, broken circuit boards and flattened metal cases shorting components to circuit ground. It's easy to bend a case back to an approximation of its original shape, but the mess inside may not be worth the trouble.

Liquid and electronics don't mix, yet people try to combine them all the time, spilling coffee, wine and soft drinks into their laptops, and dropping their cameras and phones into the ocean and swimming pools. Good luck trying to save such items. Now and then you can wash them out with distilled water, let them dry for a long time, and wind up with a functional product. Most of the time, and especially with salt-water intrusion, it's a total loss.

Just being near salt water will destroy electronics after awhile. Two-way radios, navigation systems, stereos and TVs kept on a boat or even in a seaside apartment get badly corroded inside, with rusted chassis; dull, damaged solder joints; and connectors that don't pass current. Very often you'll see crusty green crud all over everything.

Speaking of the ocean, the beach is a prime killing ground for cameras. Most digital cameras feature lenses that extend when the camera is powered on. Any sand in the cracks between lens sections will work its way into the extending mechanism and freeze that baby up, and it is very hard to get all the grit out. In a typical case, the camera is dropped lens-first into the sand, and a great deal of it gets inside. I've taken a few apart and disassembled the lens assemblies, cleaned half a beach out of them, and still had little luck restoring their operation. There's always a few grains of sand somewhere deep in those nylon gears, where you can't find them, and even one grain can stop the whole works.

The Great Capacitor Scandal

Around 1990, a worker at an Asian capacitor plant stole the company's formula, fled to Taiwan, and opened his own manufacturing plant, cranking out millions of surface-mount electrolytic capacitors that found their way into countless consumer products from the major makers we all know and love. A few other Taiwanese capacitor makers copied the formula too.

Alas, that formula contained an error that caused the electrolyte in those caps to break down and release hydrogen. Over a few years, the caps swelled and burst their rubber seals, releasing corrosive electrolyte onto the products' circuit boards, severely damaging them and ruining the units.

This ugly little secret didn't become well known for quite awhile, until long after the warranty periods were expired. Billions of dollars' worth of camcorders and other costly small products were lost, all at their owners' expense. Any attempt at having repairs made was met with a diagnosis of "liquid damage—unrepairable." The disaster was so pervasive, and took long enough to show up, that many companies insisted the failures were random and have never to this day admitted any liability for the lost value.

More recently, similar electrolyte problems have continued to plague computer motherboards and the power supplies of various products, affecting even their full-sized capacitors with leads. Caps are dying after just a year or two of use. The higher heat of lead-free soldering also may be contributing to early capacitor failure.

Lawsuits have been filed, and remedial action has been taken by some manufacturers to purge their product lines of the offending parts. Still, it is highly likely you will run into bulging capacitors in your repair work, perhaps more than any other single cause of failure. Even when they're not bulging, the caps may lose their ability to store energy, showing almost no capacitance on a capacitance meter.

History Lessons

A good doctor understands the value of taking the patient's history before performing an examination. Knowing the factors leading up to the complaint can be very valuable in assessing the cause. How old are you? Do you smoke? Drink? Have a family history of this illness? What were you doing when symptoms appeared?

If you have access to a machine's history, it can provide the same kinds of helpful hints, often leading you to a preliminary diagnosis before you even try to turn it on. Here are some factors worth considering before the initial evaluation:

- *Who made it?* As discussed earlier, products from specific companies can have frequently occurring problems due to design and manufacturing philosophy. Becoming familiar with those differences may help guide you to likely issues, especially if you've seen the problem before in another unit, even of a different model, from the same maker.

  It pays to check the Internet for reports of similar troubles with the same model product. You may save many hours of wheel reinvention by discovering that others are complaining about the same failure. You might find the cure, too.

- *How old is it?* If made before the 1990s, it shouldn't have the leaking capacitor problem. It could have a lot of wear, though, with breakdowns related to plenty of hours of use. If it *was* made in the '90s or more recently, those caps are a prime suspect.

- *Has it been abused?* Dropped? Dunked? Spilled into? Sat on? Left on the dashboard of a car in the summer? Used at the beach? Had batteries in it for months or years? Had a disc or tape stuck in it, and somebody tried to tear it out? Been in a thunderstorm? Through the washing machine? Kept on a boat? Played with by kids? Cranked up at maximum volume in a club for long periods?

  Each of these conditions can lead you down the diagnosis path. A stereo amplifier used gently at home by 70-year-olds is likely to have a very different failure than one cranked up to high volume levels in a club, or one run 40 hours a week in a restaurant for 10 years.

- *What was it doing when it failed?* While gadgets sometimes quit while in operation, many stop working when sitting idle, and the problem isn't discovered until the next time someone tries to use the product. This is particularly true of AC-powered machines that, like most things today, have remote controls. To sense and interpret the turn-on signal from the remote, at least some of the circuitry has to be kept active at all times. VCRs, DVRs, DVD players and TVs are never truly turned off; some power always flows. A power surge, a quick spike, or perhaps just age or—as always—bad capacitors can kill the standby supply, resulting in complete loss of operation.

If the device did crash while being used, it's very helpful to know precisely what operation was being carried out when it quit. If a laptop's backlight went dead while the screen was being tilted, for example, that's a good indication of a broken internal cable, rather than a blown transistor in the backlight inverter.

- *Did it do something weird shortly before quitting?* Many failing circuits exhibit odd operation for anywhere from minutes to seconds before they shut down altogether. That peculiar behavior can contain clues to the cause of their demise. In fact, it usually does, and it may hold the only hints you have in cases of total loss of function.
- *Was it sudden or gradual?* Some causes of failure, such as drifting alignment, dirty or worn mechanisms, and leaking or drying electrolytic capacitors, may manifest gradually over time. Bad caps on computer motherboards are a great example of this, as they cause the machine to get less and less stable, with more and more frequent crashes, until bootup is no longer possible.

Parts don't blow gradually, though. While it's possible in rare cases for components, and especially transistors, to exhibit intermittent bad behavior, a truly blown (open-circuited) component goes suddenly and permanently, frequently shorting first and then opening a moment later from the heat of all the current passing through its short. So, if the symptoms appeared gradually, it's a safe bet that the problem is *not* blown parts.

Stick Out Your USB Port and Say "Ahhh": Initial Evaluation

Before you take a unit apart, examine it externally and try to form a hypothesis describing its failure. The most potent paintbrush in the diagnostic art is simple logic. Your first brushstroke should be to reduce variables and eliminate as many areas of the circuitry from consideration as you can. Instead of chasing what might be wrong, first focus on what the problem *can't* be. By doing so, you'll sidestep hours of signal tracing and frustration. Before you open the unit, give some thought to these issues.

It's dead, Jim! "Dead" is a word many people use when something doesn't work, but often it's incorrectly applied. If *anything at all* happens when you apply power, the thing isn't dead! A lit LED, a display with something—even something scrambled and meaningless—on it, a hum, a hiss, some warmth, or any activity whatsoever, indicates that the circuitry is getting some power from the power supply, at least. "Dead" means *dead*. Zip, nada, nothing, stone cold. If you do see signs of life, some power supply voltage could still be missing or far from its correct value, but the supply is less likely to be the problem. In a product with a switching supply, you can assume that the chopper transistor is good, as are the fuse and the bridge rectifier. You *can't* be sure the supply has no other problem like bad capacitors or poor voltage regulation.

If the device is totally dead, check the fuse. All AC-powered products have fuses, and so do most battery-operated gadgets, though their fuses may be tiny and soldered to the board. A blown fuse pretty much always means a short somewhere inside, so don't expect much merely by changing the fuse. Most likely, it'll blow again immediately. Still, give it a try, just in case. Be sure to use the same amperage rating for your new fuse; using a bigger one is asking for trouble in the form of excessive current draw and more cooked parts, and a smaller one may blow even if the circuit is working fine. And no matter how tempting it might be, do *not* bypass the fuse, or you will almost certainly do much more damage to the circuitry than already exists. Those fuses are there for a reason, and that reason is protection.

Though nontechnical types tend to think that truly dead machines are the most badly damaged and least worth fixing, the opposite is usually true. Total loss of activity typically indicates a power supply failure or a shorted part that has blown the fuse. In other words, easy pickings. The really tricky cases are the ones where the thing almost works right, but not quite, or it works fine sometimes and malfunctions only if you turn it facing south during a full moon on a Tuesday. Those are the unruly beasts that may cause you to emit words you don't want your kids to hear.

If the product has a display, is there anything on it?   Although a dead display can be caused by many things, the condition usually indicates that the microprocessor at the heart of the digital control system isn't running. Micros rarely fail, except in cases of electrical abuse like lightning strikes or severe static electricity. The most frequent reasons for a stopped microprocessor are lack of proper power supply or a clock crystal that isn't oscillating.

If the display is there but isn't normal, that's a sign that some other issue in the digital system is scrambling the data going to it. If it's a simple system in which the microprocessor directly drives the display, the micro still might be stopped or damaged. If there's a display driver chip between the micro and the LCD, it may be bad. When the unit responds to commands but has a scrambled display, the micro is probably okay. If everything is locked up, suspect the micro or its support circuitry.

Does it work when cold and then quit after it warms up?   Thermal behavior can be caused by bad solder joints, flaky semiconductors and bad capacitors. It usually manifests as failure after warm-up, but now and then it's the other way around, with proper operation commencing only *after* the unit has been on for awhile. Again, the problem is not a blown part.

Does tapping on it affect its operation?   If so, there's a poor connection somewhere. Typically it's a cold solder joint or an oxidized connector. Cracked circuit board traces used to be fairly common, but they're quite unusual now, except in cases of physical abuse. Faulty conductive-glue layer interconnects can make boards tap-sensitive. On *very* rare occasions, the bad connection may be inside a transistor, and I once found one inside an intermediate frequency (IF) transformer in a radio receiver.

Eliminating variables   If the device runs off an AC adapter, try substituting your bench power supply, being careful of polarity, as discussed in Chapter 3. If the unit can operate from batteries, put some in and see what happens. The remote control won't turn it on? Try using the front panel buttons. Even if these attempts don't restore operation, at least you'll know what *isn't* causing the trouble.

Speaking of remotes, they can go wild and emit continuous commands, driving the micro in the product out of its little silicon mind and locking out all other attempts at operation. The situation usually occurs when liquid has been spilled on the remote, causing one or more of the keys to short out. The remote thinks a key is being pressed and sends data ad infinitum. To be sure that isn't the problem, remove the batteries from the remote and see if the symptoms disappear.

Use Your Noodle

Once you've tried these preliminary experiments, think logically about their results and you will probably have a pretty good sense of where to poke your scope probe first. Let's look at some real-life cases from beginning to end, and how this approach helped get me started in the right direction.

Stereoless Receiver

The unit was a fairly high-end stereo receiver with a dead left channel that nobody in the shop could bring back to life. Eventually they'd given up, and the set had languished on the shelf for two years by the time it and I met. The shop's owner handed it to me as an employment test. If I could fix *that* one, I was in. The smug look on his face told me I was in for a challenge.

I saw no evidence of obvious damage or abuse, so I hooked up a pair of speakers, connected a CD player for a signal source, and fired it up. My initial evaluation was that the power supply had to be okay because the right channel worked fine. The front panel lit up and the unit seemed to operate normally, other than its having a stubborn case of mono. I hooked a clip lead to the antenna terminal and tried FM reception, thinking that the trouble might be in the input switching circuitry feeding audio from the input jacks to the amplifier stages. Nope, FM sounded great, but still from only one channel.

There was no hum in either channel's output, so the power supply wasn't being bogged down by a short someplace. (A loudly humming channel with no audio is classically indicative of a shorted output transistor.) I plugged in headphones, because sometimes amplifiers with bad output stages can drive a little bit of distorted signal into headphones, though they can't power a speaker. I kept the cans off my ears, as always, just in case the thing blasted me with punishing volume. There was no difference this time; I couldn't hear a hint of audio from the bad channel, even with the balance control turned all the way to that side. It was as quiet as a mouse. A dead mouse.

I'd eliminated as many variables as I could, so it was time to open 'er up. Several techs had taken their best shots at the poor thing, and evidence of their endeavors was all over the inside. The output transistors had been changed and large components in the power supply resoldered. Other solder work indicated that resistors in and near the bad channel's output stage had been pulled and tested. The focus clearly had been toward the output stage, which very often dies in audio amps and is where most techs look first. It made sense, but it hadn't done any good this time.

Thanks to the working channel, I didn't head straight to the power supply. Since the other guys had replaced the output transistors, I didn't bother to check those either. Instead, I stuck my scope probe on the signal line feeding the output stage, and there was no audio signal. Thus, the trouble was farther back in the chain toward the input stages someplace, and everybody had been hunting in the wrong place!

I looked closely at a few small-signal transistors and traced their connections between stages. Some amplifiers are capacitively coupled (there's a capacitor between each stage), while others are directly or resistively coupled. The direct and resistive styles are also called DC coupling, because the voltages on one stage get passed to the next. It's a tougher type of circuit to design, but it results in superior sound. So most good audio gear works that way, and I expected to see that kind of circuitry here.

I wasn't disappointed; this baby had resistors between stages, but no capacitors. Thus, the DC voltage levels on one stage could affect those on the succeeding stages. A little light was beginning to glow in the back of my mind, but I needed to take a few measurements before coming to any conclusions.

I went all the way back to the first stage, finding it by tracing the line from the input jack, through the selector switches and to the amplifier board. I had a known good channel to use as a reference, so I fed the same audio signal to both sides, using a "Y" adapter cable. Setting the scope for dual trace display and the same voltage range on both input channels, I compared the outputs from the receiver's first left and right channel stages. They looked identical. Same signal levels, same DC voltage. I went to the next stage. The good channel showed 1-volt DC at that stage's output, while the bad side only had about 0.5 volts, with the same audio signal riding on both. Hmmm...could such a small difference matter? Half a crummy volt? In a DC-coupled amplifier, you bet it could! Transistors need a "bias," which is a little bit of DC to keep them turned on, at their bases (input terminals), and not having a high enough bias will make them cut off, unable to pass any signal. I checked the next stage in the bad channel, and its output was dead, just a sad, flat line on my scope. Without proper bias, the stage was completely cut off. There was the trouble! But why?

I went back to the stage with lower DC output and checked the voltages and signals on the transistor's other terminals. They matched those of the good channel. Only the output was different. So, most likely, the transistor was just dropping too much voltage. In other words, a bad transistor. A whole 25 cents' worth of mysterious mischief that had stymied an entire shop, simply because it wasn't the usual problem. I popped in a new transistor, and voilà! The entire channel came to life and worked perfectly. Just to be sure all was well, I checked the previously dead stage's output levels, and both the signal and DC level matched those of the good channel. Case closed. I got a few open-mouthed stares from the other techs over that repair, along with an offer of full-time employment at the shop. I decided not to work there, but the episode left me feeling like Sherlock Holmes solving a perplexing crime. All I needed was a pipe and an English accent. "Elementary, my dear Watt-son!"

Silent Shortwave

A friend brought me this set after buying it for very little, knowing it didn't work but badly wanting it to, as he'd always longed for one of these models and they were hard to find. One of the better digitally tuned shortwave receivers, this portable radio had no reception at all. It wasn't dead, though; the display came up normally, and a little hiss came from the speaker. Logic brain, spring into action! Where should I start?

The first thing I did was try the various bands. AM, nada. Shortwave, same. FM... hey, the FM worked! Sounded great. The FM band is at a much higher frequency and uses a different kind of signal than AM and shortwave (which is also AM), so all multiband radios have separate stages dedicated to FM reception, and clearly they were fine. The audio and some other stages are shared, though, so the working FM also confirmed that the power regulation, digital control and audio stages were all functioning. Thus, the trouble had to be in the RF (radio frequency) or IF (intermediate frequency) stages of the shortwave section, which also handled AM, or in the digital frequency synthesizer that controlled the tuning.

I discounted the frequency synthesizer because the FM worked. There still could have been trouble there, but it wasn't suspect number one. Let's see, the synthesizer generates the oscillator signal that mixes with the incoming radio signal from the antenna, resulting in the IF signal, which is then amplified by the IF stages. Then, in good radios like this set, that signal is mixed with yet another oscillator, this one of fixed frequency, to create a lower-frequency IF signal that passes through yet more amplifier stages before it is demodulated into audio.

The trouble could have been anywhere along that chain, but experience reminded me to check that the fixed oscillator, called the *second local oscillator*, was actually running. Back in the '70s, when I'd worked in the service department of a large consumer electronics chain, tons of CB radios had come in with dead receivers, thanks to a bad batch of oscillator crystals. We'd change 'em and be done in a jiffy, fixing the units without even having to troubleshoot them, since they always had the same problem. I did so many of them that the issue of a dead second local became permanently embedded in my mind. I looked for this set's crystal and touched each end with my scope probe, checking for a nice sine wave of a few volts. Nothing. The oscillator was *not* running. Aha!

Sometimes weak crystals can be jolted into operating by adding some capacitance to one end, increasing the voltage drop across the crystal because of the extra load and making it vibrate a little harder. So, I touched my finger to each lead of the crystal, one lead at a time, with another finger touching circuit ground via a metal shield, employing my hand as a capacitor. This was all very low-voltage, battery-operated stuff, and it was safe to do that. The first try, nothing. The second, wham! The radio sprang to life and the BBC boomed in loud and clear from thousands of miles away. I let go and silence filled the room again. Ah, a bad crystal, and this one would have to be ordered from Japan. I tried resoldering it, just in case it had a cold joint. No luck. Then, glancing at my friend's glum expression of disappointment that a new crystal would have to be procured from halfway around the world—restoration of the radio would be months away—I decided to grab a magnifying glass and take a close look at the surrounding components. I spied a tiny surface-mount capacitor connected from one end of the crystal to ground, performing essentially the same function my finger had. The solder joint on that one looked awfully dull. I resoldered it and the radio starting playing its little heart out. "This is London calling. And now the news...." Cost: zero. Grin on elated friend's face: priceless.

The Pooped Projector

How about a nice, high-resolution DLP video projector, with plenty of lamp life left, for $20? Sure, we'd all go for that, right? Oh, there's one small catch: it doesn't work!

I snapped up this craigslist puppy because I knew from the history of its failure exactly what was wrong before I ever saw it. The owner told me that it had started turning itself off randomly and becoming difficult to turn on. Eventually it stopped responding altogether. Now what could possibly cause that? Obviously, it couldn't be a blown part. You guessed it: a classic electrolytic capacitor failure. I could picture just what it would look like with its bulging top. I figured it'd be at the output of the internal switching power supply, probably near the DC output end of the board.

Got it home, opened it up, and there was the cap, precisely as I'd pictured it, bulge and all. It was even *where* I'd expected it. I changed the part with an exact replacement I found on one of my scrap boards, a power supply from a computer. Fired up the projector and she was good to go, with a sharp, bright picture.

While I gave it the bench test, I checked online and found numerous complaints of the same problem in this model, along with various lay diagnoses, including some wacky guesses and the correct answer. The design kept the power supply turned on at all times, stressing that particular cap and causing it to fail after a couple of years, regardless of how much use the projector got. I keep my unit unplugged when I'm not running it, so it should last for a long, long time.

How can you beat a $20 video projector? And *that*, gentle reader, is why repairing electronics is not just fun, it's incredibly economical.

Chatterbox DVD Player

This portable DVD player came from the carcass pile at a repair facility for which I worked part time. The machine, one of the better brands, had been a warranty claim, and nobody could fix the thing, so it had been replaced and kept for parts. With its 5-inch widescreen LCD, the player looked kinda cute, and it seemed a shame to cut it up. The shop's owner didn't care if I took it home, so I did. I had no idea what might be wrong with it, but the price was right.

It appeared intact, so I hooked up my bench supply and flipped on the juice. The screen lit up and the mechanism immediately started making a noise like a machine gun! I killed the power in a hurry, because I knew what that rapid-fire sound was.

Disc players use leaf switches to sense when the laser head has returned fully to the initial position at the inside of the disc, where it needs to go to begin the start-up sequence leading to disc playback. The "rat-a-tat" noise was a clear indication that the micro didn't know the mechanical limit had been reached. The unit was cranking its sled motor indefinitely, grinding the nylon gears against each other until they slipped, over and over. I could just imagine the toothless mess it might make of those delicate plastic parts if I let it run for very long. Yikes.

On opening the player up, I looked for the typical leaf switch assembly and couldn't find one! Did this model use optical sensing? There was no trace of that either. I gently turned the sled motor's gear and moved the head away from the starting position, but I still couldn't see a switch. Finally, I removed the entire spindle assembly, and there it was, a tiny leaf switch hidden underneath the disc motor. It looked fine, though. Why wasn't it being tripped? Or maybe it was, and its signal wasn't getting back to the micro for some reason. Or perhaps the micro was bad....

I forced myself off the trail of wild imagination and back onto a path of pursuit. The simplest explanation was that the switch must not be getting pressed far enough to work. I disconnected one wire from the leaf switch and connected my DMM across it, watching for the resistance to change from infinite (an open circuit) to near-zero (a closed one) as I slowly turned the gear to move the head back toward the switch. The head hit its mechanical limit and would go no farther, but the switch never closed. That was the problem, all right.

After moving the head away again, I could see why, and it was so silly that I couldn't imagine why nobody had caught it. The little metal arm on the laser head that pressed on the switch was bent—not a lot, but just enough to keep it from pushing the leaf far enough to contact its mate. I bent the arm back ever so slightly, and I had a DVD player! Almost. Alas, the disc spindle assembly's three mounting screws also worked to set the disc alignment perpendicular to the optical head as it traversed the radius of the disc, and I'd had to unscrew them to remove the spindle. Any significant tilt would cause the reflected laser beam to miss the center of the head's lens, resulting in poor tracking and skipping. And, with the alignment scrambled, it did. I found the proper test point to use for observing the head's output signal (we'll explore how to do that in Chapter 14), scoped it and redid the alignment, carefully adjusting those three screws until I got a good signal no matter what part of the disc I played. Making me mess up a critical alignment to reach the leaf switch—talk about poor design!

I won't mention the manufacturer's name, but I'd seen flimsy metal parts in some of their other products, so it wasn't terribly surprising to find one here too. This particular player went on to develop a baffling, chronic problem with the ribbon cable going to its disc motor, causing it to fail to spin the disc fast enough, resulting in an error message and no playback. I kept cleaning the ribbon's contacts and reseating the connector at the circuit board end, and it would work for a few months before failing again. Finally, I checked the other end of the cable, which had looked okay, and that was the real trouble; I'd just been wiggling it a little while working on the wrong end, and the movement had helped its connection for a short time. I cleaned and reseated that end, and the unit works to this day. Another mystery solved, another lesson learned in never assuming anything, and another fun freebie.

Chapter 5

Naming Names: Important Terms, Concepts and Building Blocks

While many different terms are used to describe electrons and their behavior, you will encounter a core set, common to all electronics, in your repair work. Some deal with electrical units, some with parts and their characteristics, some with circuit concepts, and others with hip tech slang. (Okay, you can stop laughing now!) Others describe frequently employed circuits used as the building blocks of many products. Getting familiar with these terms is crucial to your understanding the rest of this book, so let's look them over before moving on to Chapter 6.

We'll touch briefly on the most vital terms here; for more, and greater detail, check out the Glossary at the back of the book. You'd be doing yourself a favor to read the entire Glossary, rather than just using it for reference. Otherwise, you'll find yourself flipping pages back and forth a great deal as you read on. And believe me, you don't want to miss the definition of *magic smoke*.

Electrical Concepts

Being an intangible essence, electrical energy must be described indirectly by its properties. It possesses quite a number of them, and many famous scientists have teased them out with clever observations throughout the last few centuries. Experiments with electricity have gone on since the 1700s, when Ben Franklin played with lightning and miraculously lived to write about it—talk about *conducting* an experiment! Volta and Ampère built batteries and watched how their mysterious output affected wires, compasses and frogs' legs. Ohm quantified electrical resistance, and many others contributed crucial insights into this amazing natural phenomenon's seemingly bizarre behavior.

Perhaps the most valuable discovery was that electricity is a two-quantity form of energy. Its total power, or ability to do something like light a lamp, spin a motor or push a speaker cone, has two parts: how much of it there is, and how strongly it pushes.

The *ampere*, or *amp* for short, is a measure of how much electrical *current* is moving through a circuit. Interestingly, actual electrons don't travel very fast at all and are not what moves through the wires, semiconductors and other parts of a device. Rather, their charge state gets transferred from atom to atom, raising the energy level of each one's own electrons, thus passing the current along. Imagine throwing a stone into a pond and watching the resulting wave. The wave propagates outward, but do the atoms of water at the center, where you threw the rock, actually wind up at the edge of the pond? No; they hardly move at all. They just push against the atoms next to them, transferring the stone's mechanical energy from one to the next.

The amount of current, or number of amperes moving through a circuit, has nothing to do with how hard they push, just as the amount of water in a hose has no relation to how much pressure is behind it. The pressure is what we call *volts*, a measure of how high each electron's energy state rises. Volts tell you how much pressure is pushing the amps through the circuit. In fact, voltage is sometimes referred to as *electrical pressure*, or *electromotive force*. It is unrelated to how much electricity there is, just as the pressure in a hose doesn't tell you how much water is present. Volts propel current through the circuit. After all, without pressure, the water will just sit in the hose, going nowhere, right?

The hose isn't infinitely large, of course, and doesn't permit a perfectly free flow. Friction opposes and limits the motion. When the "hose" is a wire, that means it has *resistance*. Resistance is basically friction at the atomic level, and the energy lost to it from electrons and atoms rubbing against each other is converted to heat. The term for resistance is *ohms*, after the man who deduced the relationship between current, voltage and resistance. We call his crucial insight into electrical behavior Ohm's Law. If you hate math and don't want to memorize formulas, at least get the hang of this rather simple one; it's the most important, useful relation in all of the electrical arts, and grasping its essence will greatly aid your troubleshooting. See *Ohm's Law* in the Glossary.

When you put voltage and current together, you get the total picture of the power of the power, so to speak. We call that *watts*, and it describes how much work the energy can do. Determining watts is simple: just multiply the volts times the amps. So, 25 volts at 4 amps equals 100 watts, and so does 5 volts at 20 amps. Either arrangement could be converted to the same amount of mechanical work or produce the same light or heat.

As it comes from a battery, electricity is in the form of *direct current*, meaning it moves only in one direction. The side of the battery with excess electron charge is called *negative*, while the side with a lack of it is called *positive*. Thus, by definition, current passes from negative to positive as it attempts to correct the imbalance of charges. Why not the other way around? I suppose we could have named either terminal whatever we wanted, but those names were known in Franklin's time and have persisted. And they relate to our modern model of the atom, with the electron's negative charges, so I doubt anyone's going to change them.

Flipping the *polarity*, or direction of current, back and forth turns out to create many useful effects, from easing long-distance power transmission to the magic of radio signal propagation. That's *alternating current*, or *AC*, and you'll see it in just about everything. How fast you flip it is the *frequency*, specified in *hertz* (Hz). The old term was *cycles per second*. It's nice to have one word for it, don't you think? I wonder why we don't have one for speed, instead of miles per hour. We should call them glorphs. "I'm sorry, sir, you were going 45 glorphs in the 25 glorph zone. License and registration, please."

When you put two conductive plates in proximity to each other and apply voltage, they talk to one another in a peculiar way. A charge builds up on either side of the insulator between them, and that charge can be taken out and turned back into current. We call that phenomenon *capacitance*, and the parts doing the job are *capacitors*. Essentially, capacitors act like little storage wells of electricity.

Electricity and magnetism are very related things, and they interact with each other. In fact, one can be turned into the other quite easily, by passing a current through a coil of wire or by moving a magnet in a coil of wire. Passing current through the coil generates a magnetic field, and moving a magnet through a coil generates current. When a coil generates a magnetic field and then the direction of applied current reverses, the field collapses on the coil and generates current in it in the opposite direction to the current that created the field. Essentially, the coil stores some energy in the magnetic field and then puts it back into the circuit, but going the other way. That behavior is called *inductance*, and it has all kinds of very important implications in alternating-current circuits. A coil used that way is an *inductor*. Two different-sized coils wound on a common metal core can be used to transform one combination of current and voltage into another, with the magnetic field created by one generating current in the other. That's a *transformer*.

The effect capacitors and inductors have on AC currents is called *reactance*, and the combination of capacitive reactance, inductive reactance and resistance is known as *impedance*. That's an especially apt term because it quantifies the amount the circuit impedes the passage of the AC current going through it. Though it doesn't behave exactly like resistance—it's frequency-dependent, for instance—impedance is similar to resistance for AC current and is specified in ohms, just like pure resistance.

Circuit Concepts

When you wire up a bunch of parts such that current can pass through them and return to its point of origin, you create a *circuit*. The circuit concept is central to all electronics, and virtually every device that does anything is part of one. So, naturally, lots of terms are used to describe the functions and characteristics of circuits and the signals that flow through them.

When two or more circuit elements (components) are wired so that the current has to pass through one of them to reach the other, they're in *series*. Examples of things in series are fuses and switches; nothing can reach the rest of the circuit without passing through them first. It makes sense that the current through each element would have to be the same, since the amount of electricity reaching the return end of the power source has to equal what left it in the first place. Indeed, that's true. The current that passes through each element of a series circuit is always equal.

Still, energy has to be used in order for the part to do anything, so something has to give. What changes is the voltage. Each element drops the voltage, essentially using up some of the electrical pressure, until the total drop equals the applied voltage. The elements don't necessarily all drop the same amount of voltage, though. As Ohm so cleverly figured out, the amount dropped is proportional to the element's resistance. If one element has 20 percent of the total resistance of the circuit, it drops 20 percent of the voltage. Another element that has 10 percent of the total resistance drops 10 percent of the voltage, and so on. They'll always add up to 100 percent, right? Thus, all the voltage will be dropped by the time the other side of the power source is reached.

When circuit elements are wired so that multiple components are connected across the power source's two terminals, they are connected in *parallel*. In this case, each one gets the full voltage because nothing is in the way to drop some of it. The amount of current passing through each part is proportional to its resistance, regardless of the other parts also connected. Basically, they have no reason to notice each other. If you measure the total current passing through a parallel circuit, it'll add up to all the currents going through each *leg*, or element. A parallel circuit's conditions are exactly opposite to those of a series circuit: the voltage is constant but the current varies.

Circuits with a path from one end of the power source to the other are said to be complete, or *closed*. That's the normal operating state; unless a circuit is closed, nothing flows and nothing happens. When there's a break in the path, perhaps from a switch in the "off" position or a blown fuse, energy flow stops and the circuit is *open*. Any failed component no longer capable of passing current is considered open as well.

A condition causing part or all of a circuit to be bypassed, so that current passes straight to the other end of the power source, is called a *short circuit*, and the parts causing the detour are said to be *shorted*. Certain types of components, especially semiconductors, often short when they fail.

Although the generation and transport system bringing power into your home provides AC, electronics really can't use the stuff. Just as you couldn't drink from a cup swinging back and forth, circuits can't take AC power and amplify or process signals with it; the changes in the power itself would show up in the output. What's needed is a nice, steady cup from which to sip. In other words, smooth DC.

Once AC is *rectified*, or converted into one polarity, it's still a series of waves of power going up and down. To smooth it into a steady voltage, some kind of reservoir needs to store some of it, so that as the wave strength approaches zero between waves, the stored energy can fill in and raise the voltage back up. That reservoir is a *filter capacitor*. It's just a big capacitor that can store enough energy to do the job, momentarily emptying itself to power the rest of the circuit until the next power wave fills it back up again.

As circuits turn on and off and their signals rise and fall, their varying use of current can pull the voltage level feeding them up and down, causing corruption of the desired signal. Smaller filters called *bypass capacitors*, placed close to the part of the circuit pulling current, store some energy to fill in the gaps, just like the big guns do. The only difference between a filter capacitor and a bypass capacitor is its size. Generally, big ones in power supplies are called filter caps, and little ones in signal processing stages are called bypass caps. Bypass capacitors are also used to provide a low-impedance AC path to ground in parts of amplifiers where that's needed, without shorting out the DC on the same connection.

The circuitry in an electronic product is not just a huge mishmash of components. It is organized into sections and, within those, *stages*. Each stage performs one function of whatever process is required for the device to do its job. A stage might be an audio preamplifier (a low-level amplifier), a tone control, a video display driver, a demodulator (something that extracts information from a signal), a position detector for a motor, and so on. At the heart of each stage are one or more *active elements*. These are the parts that actually do the work and are generally defined as being capable of providing *gain*, which you'll read about in just a few paragraphs. Supporting the active elements are passive components like resistors, capacitors and inductors. Those can alter a signal, but they can't amplify it. Without them, though, the active elements can't do their jobs.

Stages feed signals to other stages, until the device finally produces whatever output is desired. The components passing the signal from one stage to the next are called *coupling* elements, and are usually capacitors, resistors or transformers.

Signal Concepts

Signals are voltages varying in strength, or *amplitude*, to convey some kind of information. *Analog* signals vary the voltage in a pattern resembling the information itself. For instance, the output of an audio amplifier looks like a graph of the original pressure waves of sound in the air that struck the microphone. A video circuit's signal has varying voltages representing the brightness of each dot on the screen, with a rather complex method of conveying color information and synchronizing the spots to the correct place in the picture. Its graph doesn't look like an image, but it's still an analog signal, with fine voltage gradations portraying the changing picture information.

The graph of a signal is called its *waveform*. Every time the waveform repeats, that's one *cycle*. The number of cycles occurring in 1 second is the waveform's *frequency*, and the amount of time each cycle takes is its *period*. Because the voltage varies over time, it is a mathematical function, meaning that its lines can't cross over themselves. Graphed from left to right, as they are on an oscilloscope, the level at each successive moment is to the right of the preceding moment's portrayal.

The purest, most basic waveform is the *sine wave* (Figure 5-1). It is the building block from which all other waves can be created, and it has no *harmonics*, or energy at frequencies that are multiples of the wave's frequency. A sine wave sounds like a pure tone, with no characteristics suggesting any particular musical instrument or tone color. In fact, no non-electronic musical instrument produces sine waves, though some registers on the flute come close. A tuning fork comes closer.

When a signal rapidly switches between all the way on and all the way off, it assumes a square shape and is called, appropriately, a *square wave* (Figure 5-2). Close enough examination will reveal that the on/off transitions aren't entirely

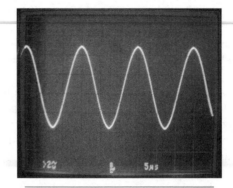

FIGURE 5-1 Sine wave

vertical, because it takes time for the state to change. Thus no square wave is truly square. The time it takes the transition to rise from 10 percent to 90 percent of its final state is the *rise time*. Going back down, it's the *fall time*.

The percentage of time spent in the "on" state, compared to the "off" state, is called the *duty cycle* and can be altered to represent information or control a motor or a voltage regulator, using a technique called *pulse-width modulation*, or *PWM*. Unlike sine waves, square waves contain harmonic energy. They include odd harmonics, but not even ones. That is, there is energy at three, five and seven times the frequency, but not at two, four and six times.

A signal used in applications requiring something to move and then quickly snap back is the *sawtooth wave*, so named for its obvious resemblance to its namesake (Figure 5-3). Oscilloscopes and CRT TVs use sawtooth waves to sweep the beam across the screen and then have it rapidly return. Other circuits, including *servos* (motor position controllers) in video tape recorders, use sawtooth waves too. Sawtooth waves include both odd and even harmonic energy.

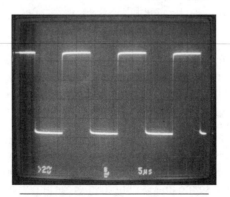

FIGURE 5-2 Square wave

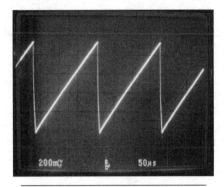

FIGURE 5-3   Sawtooth wave

The relative position in time of two waveforms is called their *phase relationship* and is expressed in degrees. As with a circle, 360 degrees represent one cycle of a waveform, regardless of how long that cycle takes. Two waveforms offset by half a cycle are 180 degrees out of phase. When there is no offset, the waveforms are in phase.

Digital signals are entirely different. Using pulses resembling square waves, digital information is always in one of two states, on or off, representing the binary numbers 1 and 0. That has tremendous advantages over the analog method, because keeping track of those two states is a lot easier than accurately moving and processing a voltage with infinitely fine gradations. Noise in a digital channel has no effect at all until it's so bad that the two states can't be found, while noise in an analog channel is very hard to separate from the desired signal and corrupts it badly. That's why scratches on an analog LP record create clicks and pops in the audio, while scratches on a CD don't. All circuits introduce some noise, so the digital method is less susceptible to degradation as it moves through various processes. Digital data is also much easier to store and manipulate, again because it has only two states to worry about.

The world of sound and light is inherently analog, though; nothing in nature exists only as ons and offs! To *digitize* natural phenomena like sounds and images, an *ADC*, or analog-to-digital converter, is used to chop the analog information into a rapid series of *samples*, or measurements, which are then encoded, one by one, into the binary 1's and 0's of digital data. Conversion is a complicated process that introduces quality limitations of its own, so digital is no more perfect than is analog. Digital's imperfections are different, though, and generally less objectionable.

Building Blocks

There's one heck of a variety of circuits out there! For any given function, a designer can find lots of ways to build something that works. While the circuitry "wheels" get reinvented all the time, they're all round and they all spin, so common circuit configurations are found in pretty much all products. Sometimes they have significant variations, but they're still basically the same old thing and can be recognized easily once you get familiar with them. Let's look at some common circuits you're likely to find.

The basic circuit at the heart of most stages is an amplifier of some kind. Amplifiers have *gain*, which means they take steady DC from the power supply and shape it into a replica of an incoming signal, only bigger. That "bigger" can be in terms of voltage, current or both. Most voltage amplifiers also flip the signal upside down, or *invert* it. Sometimes that's important to the circuit's operation, but much of the time it's just an irrelevant consequence of how voltage amplifier stages are configured.

A very common type of current amplifier is the *emitter follower*. This one takes its output from the emitter of a transistor, and the amplified signal mimics, or follows, the input signal, without inversion or change in voltage swing. Only the amount of current the signal can pump into a *load* is increased.

When amplifying analog signals, *linearity*, the ability to mimic the changes in the input signal faithfully without distortion, is a critical design parameter. The term comes from the graph that results if you plot the input signal against the amplified output. In a truly linear circuit, you get a straight line. The more gain, the more the line points upward, but it's still straight, indicating a ratio of input to output that doesn't change as the signal's voltage wiggles up and down. If the amplifier is driven past the point that its output reaches the power supply voltages, the transistors will be all the way on or all the way off during signal extremes, resulting in an output that no longer accurately follows the input. That's serious nonlinearity, also known as *clipping* because it clips off the tops and bottoms of the waveform; the amplifier simply can't go any farther. If you've ever turned up a stereo loud enough to hear ugly distortion, you've experienced clipping. Even a small amplifier driven to clipping can burn out the tweeters in a pair of speakers that normally could take the full power of a bigger amplifier without harm. The high-frequency content of a clipped waveform is much greater than that of a linearly amplified signal, thanks to the steep edges of the clipped area, so it drives disproportionate power into the tweeters. I've seen it happen. It can injure your ears, too.

Most high-fidelity audio amplifiers, and even many small ones of the sort used in pocket radios and MP3 players, use a *complementary* design, referred to as Class AB. Complementary amplifiers split the audio waveform into its negative-going and positive-going halves and amplify each half separately. Then they combine the two halves at the output, rebuilding the waveform. Why do that? The technique allows for excellent efficiency because almost no power is dissipated when the waveform is near the zero voltage level. Only when the input signal gets big does the amplifier draw a lot of power, keeping power usage proportional to output.

In a non-complementary design, the amplifier has to be *biased* to set its output halfway between ground and the power supply voltage when no signal is applied, so that the negative and positive peaks of signals can make the output swing both higher and lower in step with them. It sits there eating half the available power at all times. Some high-end amplifiers, called Class A, actually do it that way to avoid certain subtle distortions associated with splitting and recombining the waveform. Those amplifiers sound great but run very hot and waste a lot of power.

Tuned amplifiers use *resonant* circuits to select a particular frequency or range of frequencies to amplify, rejecting others. They're used in radio and TV receivers to separate and boost incoming signals. See *resonance* in the Glossary. The very first amplifiers in a receiver that strengthen the weak signals from the antenna are called the *front end*. Some are tuned, and some aren't, depending on the design. Later amplifier stages, operating at a fixed frequency to which all incoming signals are converted, are called *intermediate frequency*, or *IF*, amplifiers. Those are always tuned, and they provide most of a receiver's *selectivity*, or ability to separate stations.

Even digital gates in integrated circuits are amplifiers. They take in digital pulses and amplify them enough to ensure that they swing all the way to the supply voltage and ground, making up for any losses that may have occurred and preventing them from accumulating until the pulses can no longer be reliably processed. These amplifiers are deliberately nonlinear; the only desired states are fully *saturated*, or all the way on, and fully *cut off*. Their *linear region*, where small changes in the input signal would be faithfully amplified, is made as narrow as possible. Such small wiggles in digital signals only represent noise anyway.

Oscillators generate their own signals. They have many uses, including providing timing pulses to, or *clocking*, microprocessors, generating tones, mixing with radio signals to convert their frequencies, and lots more. An oscillator is basically an amplifier with its output fed back to its input in phase, reinforcing the input and sending the signal around and around again indefinitely.

Oscillators can be designed to produce any of the basic waveforms. In analog signal processing circuits, sine wave oscillators are often the most useful, thanks to their purity. Sawtooth wave oscillators are used to drive electron beams across CRTs and any time something needs to be swept and then quickly returned to its starting point. With digital circuits, which operate with pulses, square waves are the order of the day.

The significant challenge with most oscillators is setting the frequency and keeping it constant. When only a single frequency is needed, a quartz crystal or ceramic resonator can keep the oscillator very accurate. These parts resonate mechanically at the molecular level, and they're dimensionally stable, so they drift very little with temperature. The tradeoff is that a given crystal can generate only one frequency.

When frequency variability is required, as in a radio tuner, simple resonant circuits like inductor/capacitor combinations work but are not terribly stable, especially regarding thermal drift.

While early radios could tolerate some drift, today's high-precision systems simply can't. Do you really want to get up to fine-tune your digital HDTV every 20 minutes as the receiver drifts off-frequency and the picture drops out? Of course not! The receiver has to sit on its tuned frequency the moment you turn it on and stay there all day long.

The *digital frequency synthesizer* solves the stability problem by providing frequency agility while still being referenced to the unvarying frequency of a quartz crystal. Pulling that off isn't simple. There are two basic techniques. In a classic hybrid analog/digital synthesizer, an analog oscillator's frequency is controlled by a voltage from the synthesizer. The resulting frequency is digitally divided or multiplied until it matches that of the crystal. The two are then compared, and the controlling voltage is adjusted to keep them at the same frequency. So, even though the analog oscillator really isn't at the same frequency as the crystal, the comparison circuit thinks it is. This kind of circuit is called a *phase-locked loop*, or *PLL*.

Tuning the oscillator is accomplished by changing the ratio of division and multiplication, forcing the control voltage to change the true frequency to match the result of the division to that of the unvarying crystal. Many receivers have been built this way, but the method has a serious downside. In order for the comparison and correction process to work, there has to be a little error to correct. The oscillator's frequency wobbles ever so slightly as it drifts and gets corrected, resulting in a phenomenon called *phase noise*. The noise can be kept low, and some very nice radios worked just fine with this style of synthesizer, but it caused enough signal distortion that better methods were sought.

As digital technology advanced, chips got fast enough that the oscillator could be done away with altogether, and the required output signal could be built directly from digital data and converted to analog, in much the same way a CD player rebuilds the audio waveform of a music disc from samples. In many applications, this *direct digital synthesis*, or *DDS*, technique has replaced hybrid analog/digital synthesizer designs. Not everywhere, though: you'll still find PLL synthesizers in lots of UHF and VHF receivers because today's chips are only now getting fast enough to create the required signals at such high frequencies. The tipoff is if you see a component on the schematic that looks like a combination of a capacitor and a diode. That's a *varactor*, or voltage-variable capacitor, and it's what tunes the analog oscillator with the digital part of the circuit's control voltage. Where there's a varactor, there's a PLL.

PLLs are used for other purposes too. Digital data is recovered from media such as hard drives and optical discs using a PLL to synchronize the data rate to the circuit detecting it. Analog VCRs and camcorders use PLLs to recover the wobbly color information from the tape and stabilize it to the very high precision required for proper display.

Servos are a lot like PLLs, except that they slave a motor's rotational speed and phase (position at a given moment) to a reference signal. A servo regulates a DVD player's disc rotation, keeping it at whatever speed is required for a given data rate. The rapid lens motions required to focus the laser beam on the microscopic pits and track them as they whiz by are also controlled by servos. In a VCR, servos adjust capstan motor and head drum rotation, locking them to a signal recorded on the tape, so the rotating heads will correctly trace over the recorded tracks.

Voltage regulators keep power supply voltages constant as current demand varies with circuit function. Linear regulators use a *series pass transistor* as a variable resistor, automatically changing the resistance to set the voltage, and dissipating the unwanted extra power as heat. They're simple and effective but also inefficient. Linear regulators handling serious power get quite hot.

Switching regulators use pulses to turn a transistor on and off like a switch, and then reconstitute the pulses passed through it back into steady DC power with a filter capacitor. Changing the pulse width permits more or less power to get through the transistor over a given period of time. Because the transistor is almost always completely on or completely off, except for the short moments when it switches states, little power is converted to heat, and there's no excess to waste, since only as much energy as needed is allowed through. Switching regulators are more complicated, and they have the potential to generate electrical noise, but their cool-running efficiency makes them very desirable, especially in battery-operated devices, where conserving power is critical.

Now that you've seen some basic terms, core concepts and circuit building blocks, go read the Glossary and see lots more. Seriously. Do it now, before going on to the next chapter. I know, "Grrr, this guy is such a nag." You'll thank me later. Really, you will. And, hey, there's some cool stuff there. You'll enjoy it, I promise.

Chapter 6

Working Your Weapons: Using Test Equipment

Effective application of test gear is key to your sleuthing success. Especially with the oscilloscope, the settings you make while probing around a circuit determine what elements of the signals you will see and which ones you may miss. Soldering, too, can be more or less effective, depending on your technique. Let's look at each piece of basic test gear and how to use it to your best advantage.

Digital Multimeter

Digital multimeters (DMMs) are great for measuring things that don't change quickly. Battery and power supply voltage, along with resistance and current, are prime candidates for being checked with a DMM. The instrument is less effective for observing changing voltages and currents, which look like moving numbers and are tough to interpret.

Overview

The DMM's great advantages over other instruments are its precision and accuracy. Even a digital scope has fairly limited resolution; you can't tell the difference between 6.1 and 6.13 volts with one very easily, if at all, and measuring resistance and current is impossible with normal scope setups.

All that detail in the meter's display can get you into trouble, though, if you take it too seriously. When interpreting a DMM's readings, keep in mind that real life never quite hits the specs. Don't expect the numbers you see to be perfect matches for specified quantities. If you're reading a power supply voltage that's supposed to be 6 volts, a reading of 6.1 probably isn't indicative of a circuit fault. The same is true of resistance; if the reading is very close, the part is most likely fine. And if the rightmost digit wanders around a little bit, that's due to normal noise levels or the digitizing noise and error inherent in any digital sampling system. Remember, when a part goes bad, it's not subtle! Real faults show readings far from the correct values.

Most DMMs run on batteries, and that's a good thing because it eliminates any ground path from the circuit you're testing back to your house's electrical system. The instrument "floats" relative to what's being tested (there's no common ground), so you can even take measurements across components when neither point is at circuit ground. If your DMM has the option for an AC adapter, don't use it. Always run your DMM on battery power. The batteries will last for hundreds of hours anyway.

DC Voltage

To check a circuit point's voltage, first you must find circuit ground. Usually, it's the metal chassis or metal shields, if there are any. *Don't* assume that heatsinks, those finned metal structures to which are attached larger transistors, voltage regulators and power-handling ICs, are connected to ground! Sometimes they are, sometimes they're not. In switching power supplies, the chopper transistor's heatsink may have several hundred volts on it. You sure don't want to hook your meter there.

In some devices, especially small ones like digital cameras, you may find no shields, and there's no metal chassis either. So where is ground? In most cases, the negative terminal of the battery will be connected to circuit ground, and you can use that. Particularly if you can trace it to a large area of copper foil on the board, it's a fairly safe bet. Also look for electrolytic capacitors in the 100 μf-and-up range with voltage ratings lower than 50 volts or so. (See Chapter 7.) Those are most likely power supply filter caps, even in battery-operated gear, and their negative terminals will be connected to ground. If you see two such identical caps close to each other, the device may have a split power supply, with both negative and positive voltages. Trace the caps' terminals and see if the negative lead from one is connected to the same point as the positive from the other. Where they meet is probably circuit ground.

If all else fails, you can use the outer rings of RCA jacks on audio and video gear. The only way to get an alligator clip to stay put on one of those jacks is to push half of it into the jack, with the other half grabbing the ground ring. It's better to use an input jack, rather than an output, so the part of the clip sticking inside can't short out an output, possibly damaging the circuitry. You can't hurt an input by shorting it to ground.

Turn on your DMM, set its selector switch to measure DC voltage, and connect its negative (black) lead with a clip lead to circuit ground, regardless of whether you intend to measure positive or negative voltage. A DMM will accept either polarity; measuring negative voltage simply adds a minus sign to the left of the displayed value.

With power applied to the circuit under test, touch the positive lead's tip to the point you want to measure, being careful not to let it slip and touch anything else. Many DMMs are autoranging and will read any voltage up to the instrument's ratings without your having to set anything else. Keep the probe in place until the reading settles down; it can take 5 or 10 seconds for the meter to step through its ranges and find the appropriate one.

If your DMM is not autoranging, start at the highest range and switch the range down until a proper reading is obtained. If you start at the lowest range and the voltage you're measuring happens to be high, you could damage the DMM.

If you see a nice, steady number somewhere in the voltage range you expect, it's safe to assume you have a valid measurement. If, however, you see a moving number at a very low voltage, you're probably just reading noise on a dead line, and you may have found a circuit problem. If you see a voltage in the proper range but it won't settle down, that indicates noise on the line, riding along with the voltage. Such a reading can suggest bad filter capacitors, but only when the point you're measuring is supposed to have a clean, stable voltage in the first place. Regulated power supply output points should be steady, but some other circuit points may carry normal signals that fool the DMM, causing jumping readings. To see what's going on with those, you'll be using your scope. Generally, electrolytic caps with one lead going to ground shouldn't have jumpy readings, since their reason for being in the circuit is to smooth out the voltage.

AC Voltage

You'll usually use this as a go/no go measurement. Is the voltage there or not? DMMs are optimized to read sine waves at the 60-hertz AC line frequency, so the reading doesn't mean much if you try to measure an audio signal or the high-frequency pulses in a switching power supply. Measurements are taken across two points, as with DC voltage, but in many circuits neither point will be at ground.

DMMs indicate AC voltage as root-mean-square (RMS), which is a little bit more than the average voltage in a sine wave when taken over an entire cycle. It's a useful way of describing how much power an AC wave will put into a resistive load, compared to DC power, but it is not a measurement of the actual total voltage swing. The RMS value is much smaller than the peak-to-peak voltage you'll see with your scope. American AC line voltage, for example, is 120V RMS and reads about 340 volts peak-to-peak on a scope. (If you want to keep your scope, don't try viewing the AC line with it unless you have an isolation transformer!)

For a sine wave, RMS is 0.3535 times the peak-to-peak value. For other waveforms, it can be quite different, as the time they spend at various percentages of their peak values varies with the shape of the wave. DMMs are calibrated to calculate RMS for sine waves, so the reading will be way off for anything else, at least with hobbyist-grade meters.

Resistance

When measuring resistance, *turn off the power to the circuit*! The battery in your DMM supplies the small voltage required to measure resistance, and any other applied power will incur negative consequences ranging from incorrect readings to a damaged DMM. In addition to removing the product's batteries or AC adapter (or unplugging it from the wall, in the case of AC-operated devices), it pays to check for DC voltage across the part you want to measure and to discharge any electrolytic caps that could be supplying voltage to the area under test.

Some resistances can be checked with the parts still connected to the circuit, but many cannot because the other parts may provide a current path, confusing the DMM and resulting in a reading lower than the correct value. For most resistance measurements, you will need to unsolder one end of the component. When one side of it goes to ground, leave that side connected and connect your DMM's negative lead to the ground point; it's just more convenient that way. When neither side is grounded, it doesn't matter which lead you disconnect.

Set the DMM to read resistance (Ω, or ohms). If it's autoranging, that's all you need do. Let it step through its ranges, and there's your answer. If it isn't autoranging, start with the *lowest* range and work your way up until you get a reading, so you won't risk applying the higher voltages required to get a reading on the upper ranges to sensitive parts. DMMs with manual ranges have an "out of range" indicator to show when the resistance being measured is higher than what that range can accept, usually in the form of the leftmost digit's blinking a "1." (If you're on too high a range, you'll see all 0's or close to it.)

With a manually ranging DMM, you can get more detail by using the lowest range possible without invoking the out-of-range indicator. For instance, if you are reading a 10-ohm resistor on the 20 KΩ (20,000 ohm) scale, you'll see 0.001. If you switch to the 200-ohm scale, you'll see 0.100 or thereabouts. If the resistor's measured value is too high by, say, 20 percent, which is a significant amount possibly indicating a bad part, it might show 0.120, critical data you'd miss by being at too high a range. Autoranging meters always use the lowest possible range, for the most detailed reading.

Resistance has no polarity, so it doesn't matter which lead you connect to which side of a resistor. If you're checking the resistance of a diode or other semiconductor, it does matter, and you must swap the leads to see which polarity has lower resistance. The essence of a semiconductor is that it conducts only in one direction, so a good one should have near-infinite resistance one way and low resistance the other. Checking semiconductors for resistance with a DMM can yield unpredictable results, though, because the applied voltage may or may not be enough to turn the semiconductor on and allow current to pass, depending upon the meter's design. There are better tests you can perform on those parts, but a reading of zero or near-zero resistance pretty definitively indicates that the component is shorted.

Continuity

Continuity simply shows whether a low-resistance path exists, and is intended as a "yes or no" answer, rather than as a measurement of the actual resistance. It's exactly like taking a resistance measurement on the lowest scale, except that many meters have a handy beeper or buzzer that sounds to indicate continuity, so you don't even have to look up. Use this test for switches and relay contacts, or to see if a wire is broken inside its insulation or a connector isn't making proper contact.

In many instances, you won't need to pull one side of the component to check continuity, as the surrounding paths will have too much resistance to fool the meter and invalidate the conclusion. There are some exceptions, however, involving items like transformers, whose coil windings may offer very little resistance and appear as a near-zero-ohm connection across the part you're trying to test. If you're not sure, pull one lead of the component. And, as with resistance measurement, make sure all power is off when you do a continuity check!

DC Current

Most DMMs can measure current in amps or milliamps. To measure current, the meter needs to be connected between (in series with) the power source and the circuit drawing the power, so that the current will pass through the meter on its way to the circuit. Thus, neither of the DMM's leads will be connected to ground. Never connect your DMM *across* (in parallel with) a power supply's output when the meter is set to read current! Nearly all the supply's current will go through the meter, and both the instrument and the power supply may be damaged. The meter, at least, probably will be.

Even with the meter properly connected, it's imperative that you not exceed its current limit or you will damage the instrument. For many small DMMs, the limit is 200 milliamps (ma), or 0.2 amps. Some offer higher ranges, with a separate terminal into which you can plug the positive test lead, extending the range to 5 or 10 amps.

In estimating a device's potential current draw, take a look at what runs it. If it's a small battery, as you might find in an MP3 player or a digital camera, current draw probably isn't more than an amp or so. For some devices, it's much less, in the range of 100–200 ma. If the unit uses an AC adapter, the adapter should have its maximum current capability printed on it somewhere, and it's safe to conclude that the product requires less than that when operating properly. Some gadgets state their maximum current requirements on the backs of their cases, too. When they do, they indicate the maximum current needed under the most demanding conditions—for example, when a disc drive spins up or a tape mechanism loads—and normal operating current should be less.

To take a current measurement, you need to break a connection and insert the meter in line between the two ends of it. Don't worry about test lead polarity; all you'll get is a minus sign next to the reading should you attach it backward. If you want to know the current consumption of an entire product, connect the meter between the positive terminal of the battery or power supply and the rest of the unit. If you want to measure the current for a particular portion of the circuitry, disconnect whatever feeds power to it and insert the meter there.

The DMM measures current by placing a low resistance between the meter's leads and measuring the voltage across it. The higher the current, the higher that voltage will rise. With a big current, the resistance can be very low, and there will still be enough voltage across it to get a reading. With smaller current, the resistance needs to be higher to obtain a significant, measurable voltage difference. Thus the higher ranges place less resistance between the power and the circuit. Start with the meter's highest range and work your way down. Using too low a range may impede the passage of current enough to affect or even prevent operation of the product. It also may heat up the DMM's internal resistor enough to blow it.

Current is perhaps the least useful measurement and consequently the one most infrequently performed. Now and then, it's great to know if excessive current is being drawn, but heat, smoke and blown fuses usually tell that story anyway. The more revealing result is when current *isn't* being drawn; that tells you some necessary path isn't there, or that the unit isn't being turned on. Especially because breaking connections to insert the meter is inconvenient, however, you won't find yourself wanting to measure current very often.

Diode Test

Some DMMs offer semiconductor junction tests, making them handy for checking diodes and certain types of transistors. The measurement is powered by the meter's battery, as with resistance measurements, but it's taken somewhat differently. Instead of seeing how many ohms of resistance a part has, you see the voltage across it. And to complete the test, you must reverse the leads and check the flow in the other direction.

Kill the power and disconnect one end of the component for this test. A good silicon diode should show around 0.6 volts in one direction and no continuity at all in the other. That lack of flow will be shown as the maximum voltage being applied, typically around 1.4 volts. (You can check your meter's open-circuit value by setting it to the diode test without connecting the leads to anything.) If you see 0 volts or near that, the part is shorted. To verify, switch the leads and you should see zero in the other direction too. If you see 1.4 volts (or whatever your meter's maximum is) in both directions, the part is open, a.k.a. blown. If the meter indicates the normal 0.6 volts in the conducting direction but also shows even a slight voltage drop the other way around, the diode is leaky and should be replaced.

Some DMMs perform capacitance, inductance, frequency and other measurements, but most don't. If yours does offer these readings, see the sections on those kinds of meters, and the principles will apply.

Oscilloscope

Back in Chapter 2, I emphasized that the scope is your friend. Now it's time to get acquainted with your new best buddy. This is the most important instrument, so learning to use it well is absolutely vital to successful repair work. There's no need to be intimidated by all those knobs and buttons; we'll go through each one and see how it helps you get the job done. Various scope makes and models lay out the controls differently, and some call them by slightly different names, but they do the same things. Once you get used to operating a scope, you can figure out how to use any model without difficulty.

While the functions of analog and digital scopes are basically the same, each type offers a few features unique to its species, along with some characteristic limitations.

Let's look at a scope's functions and operation, using an analog instrument as an example, with digital-specific differences noted along the way. Then we'll review some important items to keep in mind when working with a digital unit.

Overview

You cannot harm your scope by misadjusting its controls, outside of possibly damaging the CRT with an extremely high brightness setting. But even that doesn't happen in an instant. So have no fear as you play with the knobs, and feel free to experiment and learn as you go. Just keep the brightness to reasonable levels and you'll be fine.

The purpose of a scope is to plot a graph of electrical signals, with horizontal motion, or *deflection*, representing time and vertical motion representing signal voltage. Various controls adjust the speed and the voltage sensitivity so you can scale a wide range of signals to fit on the display. Others help the scope trigger, or begin drawing its graph, on a specific point in the signal, for a stable image. Still more let you perform special tricks helpful in viewing complex signals. Now you know why a scope has so darned many knobs!

To get signals on the screen, you will first connect the probe's ground clip to circuit ground of the device you're examining. If circuit ground is connected to the wall plug's round ground terminal, that's fine, but *never* connect the clip to any unisolated voltages—that is, points connected to the AC line's hot or neutral wires. Doing so presents a serious shock hazard, along with the distinct possibility of destroying your scope. (This issue crops up mostly when you're working on switching power supplies, so read up on them in Chapter 14 carefully before you try to connect your probe to one.)

Next, you'll touch the probe's tip to the circuit point whose signal you want to see, and set the vertical, horizontal and trigger controls to scale the signal to fit on the display and keep it steady. Really, that's all there is to it. The rest is just details.

The boxes on the face of the screen, called the *graticule*, are used for visual estimation of the signal's voltage and time parameters, and the vertical and horizontal controls are calibrated in *divs*, for divisions. One box equals one division, and the boxes are subdivided into five equal parts, for an easily visible resolution of 0.2 divisions, or 0.1 divisions if you want to count the spaces in between the subdivision lines.

So, if the vertical input control is set to 0.5 volts/division, and your signal occupies two divisions from top to bottom, it's a 1-volt signal. Similarly, if the time/div control is set to 0.5 μs (microseconds, or millionths of a second) and one cycle of your signal occupies two divisions from left to right, it has a period of 1 μs and is repeating at a rate of about 1 Mhz. (1/period = frequency. Something that occurs every millionth of a second happens a million times a second, right?)

Notice I said "*about* 1 Mhz." Keep in mind that scopes are not intended for measurements requiring tremendous precision or accuracy. A few percent is the best they do. Newer designs offer more accurate time calibration, thanks to digital generation of their internal timing clocks, but the precision is still low compared to that of a digital frequency counter's many digits. Plus, the vertical input specs don't approach those of the horizontal, even on digital scopes, because scaling the incoming voltage so it can be digitized is still an analog process, with all the attendant drift and error inherent in those.

Though the layout of scopes varies by manufacturer, vertical controls are usually near each other, with horizontal ones grouped together somewhere else. Typically, vertical stuff is on the left and horizontal is on the right. Other functions, like triggering and screen controls, may be anywhere, although screen settings are usually under the screen, with triggering near the far right edge of the control panel.

To begin, find the power button, most likely near the bottom of the screen. Turn the scope on and adjust the following controls.

Screen Settings

This group of controls is pretty much always located beneath the screen on CRT-type scopes. Most scopes with CRTs are analog, but some early digitals use them too.

- **Brightness or intensity**   Set it to midrange. If it seems rather bright, turn it down a bit. If you see nothing on the screen, *don't* turn the brightness way up. Other controls may need to be set before you'll see a line, or *trace*, on the display. If your scope has dual brightness controls for A and B, use A. The B control is for delayed sweep operation, which we will explore a little later on.
- **Focus**   If you can already see a trace on the screen, adjust the focus for the sharpest line. Otherwise, set it to midrange.
- **Astigmatism or astig**   Set it to midrange. Some scopes don't have an astig control, but most do.

Vertical Settings

- **Vertical mode**   This control or set of buttons may be anywhere on the scope, but it's usually near the vertical channels' controls. Set it to channel 1.
- **Channel 1's volts/div or attenuator knob**   Set it to 0.5 volts. Make sure its center knob is fully clockwise. On most scopes, it'll click at that position.
- **Input coupling or AC/DC/GND**   Set it to DC.
- **Channel 1's vertical position**   Set it to about one-third of the way up.

Horizontal Settings

- **Sweep rate or time/div**   Located to the right of the screen, this will probably be the biggest knob on the scope, and it may have a smaller knob inside, with an even smaller one in the center of that. Set the *outer* knob to 2 ms. Make sure the innermost knob at the center of this control is all the way clockwise.
- **Sweep mode**   Look for auto, normal and single. Set it to auto.
- **Horizontal display**   Look for a knob or buttons labeled A, A intens B, and B. Set it to A.

Trigger Settings

To find the trigger controls, look for a knob labeled Level. Also look for switches for coupling, source and slope. If your scope has separate trigger sections labeled A and B, use A.

- **Source**   Channel 1
- **Coupling**   AC
- **Level**   Midrange; on most scopes, the knob's indicator line will be straight up.

These settings should result in your seeing a horizontal line across the screen. If not, turn the Channel 1 vertical position knob back and forth. You should see the line moving up and down.

If you still don't see anything, try turning up the brightness control pretty far. Still nothing? Turn it back down to medium. If you saw a spot on the left side when you turned it up, then your scope is not sweeping across the screen. Check that the sweep mode is set to auto. If it's on normal or single, you will not see anything when no signal is applied. If you still have a blank screen, either you have made an error in this initial setup, your scope is not getting power, or it's not functional. Go back and check all the settings again.

Viewing a Real Signal

Assuming you do see the line, connect a scope probe to the Channel 1 (or A) vertical input by pressing its connector onto that channel's input jack and then turning the sleeve clockwise about a quarter turn until it locks. If the probe has a little 10X/1X switch on it, set it to 10X. Usually, the switch is on the part of the probe you hold, but some types have the switch on the connector, at the scope end. Touch your finger to the probe's tip and you should see about one cycle of AC on the screen. It might wobble back and forth a little, but it should be fairly stable. If all you see is a blur, try adjusting the trigger level knob back and forth until the image locks. You can also step the vertical attenuator's range up and down so the image occupies most of the screen. You're looking at the voltage induced into your body from nearby power wiring! Pretty startling, isn't it?

What All Those Knobs Do

Now that you have the scope running, let's look at what each control does and how to use it.

Screen Controls

Screen controls are specific to CRT displays. They adjust the electron beam for optimal tracing, accounting for changes due to drift, writing speed, and so on. You won't find screen controls on scopes with LCDs.

Brightness or Intensity   This sets the brightness of the trace. It should be adjusted to a medium value. Don't crank the brightness way up for very long or you may burn the trace into the tube's phosphors, and there's no undoing that. Depending on the speed of the signal you're examining, you might have to turn the brightness up so high that the trace will be way too bright when you remove the signal or slow the scope's sweep rate back down, and you'll have to back off the brightness again. Be sure to do so without much delay.

You may find two brightness controls: one for the main sweep and another for the delayed sweep. The extra brightness control is there because the beam will be sweeping at two different speeds, and the faster sweep may be too dim to see without a brightness boost.

If you turn the brightness up very high to see a fast signal, the beam's shape may distort or go out of focus a bit, even if the displayed brightness remains low. That's normal and is nothing to worry about, but it means the scope's circuits are being driven to their maximum levels, so it's a good idea to keep the brightness down below the point at which distortion becomes significant.

Focus   This focuses the beam. Turn it for the sharpest trace. When adjusting the astigmatism control, you may have to alternate adjusting astig and focus for maximum sharpness.

Astigmatism or Astig   This sets the beam's shape and should be adjusted for the sharpest, thinnest trace when displaying an actual signal. You can't set it by observing a flat line. An easy way to adjust it is to touch the probe's tip to the scope's *cal* (calibrator) terminal, which outputs a square wave signal useful for calibrating several of the instrument's parameters. You can let the probe's ground wire hang, since it's already grounded to the scope through the cable. Adjust the Channel 1 input attenuator, the A trigger level and horizontal time/div control to get a few cycles of the square wave on the screen. Turn the astig control until the waveform looks sharpest. Normally, you won't have to mess with it again. Now and then, you might touch it up when viewing very fast signals with the brightness control cranked up. If you do, you'll need to reset it afterward.

Rotation   This is usually a recessed control under the screen, accessible with a screwdriver. It compensates for ambient magnetic fields that may cause the trace to be tilted. Get a flat line, use the Channel 1 vertical position knob to center it right down the middle, and adjust the rotation to remove tilt. Unless you take the scope to another locale, you'll probably never have to touch this control again.

Illumination   This adjusts the brightness of some small incandescent bulbs around the edge of the screen so you can see the graticule better, especially when shooting photos of the screen. On a used scope, if the control does nothing, the lamps may be burned out. Their functionality has no effect on the operation of the scope. I always leave mine turned off anyway.

Beam Finder   Activating this button stops the sweep and puts a defocused blob on the screen so you can figure out where the beam went, should it disappear. If the blob is toward the bottom of the screen, the trace has gone off at the bottom. If it appears near the top, the trace is above the top. If it's in the middle, the trace is within normal viewing limits but the horizontal sweep is not being triggered to move the beam across the screen.

On some scopes, the sweep continues to run but all dimensions get smaller and everything gets brighter, so you can see a miniature version of what might otherwise be off the screen.

Vertical Controls

The vertical controls scale the incoming signal to fit on the screen. They also allow you to align the image to marks on the graticule for measurement purposes.

Probe Compensation   This makes the probe match the input channel's characteristics to ensure accurate representation of incoming signals. It is also a screwdriver adjustment, but it's not on the scope itself. Instead, you'll find it on the probe, and it can be at either end. It won't be marked, so look for a hole with a little slotted screw. Make sure the Channel 1 input coupling is set to DC. If the probe has a 10X/1X switch, set it to 10X. Touch the probe to the cal terminal, and adjust the input attenuator so that the square wave uses up about two-thirds of the vertical space on the screen. Set the time/div control so you can see between two and five cycles of the square wave. Look at the leading edge of the waveform, where the vertical line takes a right turn and goes horizontal at the top of each square. Adjust the probe compensation for the squarest shape. In one direction, it'll make a little peak that sticks up above the rest of the waveform. In the other, it'll round off the corner. It may not be possible to get a perfect square, but the closer you can get, the better. See Figures 6-1 and 6-2.

Once the probe is matched to the input channel, it'll have to be recalibrated if you want to use it on the other channel or on another scope. This takes only a few seconds.

Vertical Input Attenuator or Volts/Div   The outer ring of this control scales the vertical size of the incoming signal so it will fit on the screen. The voltage marking on each range refers to how many volts it will take to move the trace up or down one

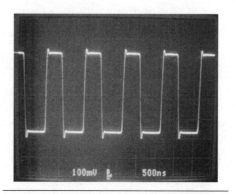

FIGURE 6-1   Probe overcompensation

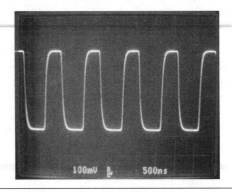

FIGURE 6-2   Probe undercompensation

division, or box, on the graticule when you're using a 1X probe, which passes the signal straight through to the scope without altering it.

Here's where things get interesting. Remember that 10X/1X switch on the probe? When set to 10X, it divides the incoming signal's voltage by ten. If your probe has no switch, it still does the same thing, as long as it's a 10X probe, which most are. You have to multiply the attenuator's setting by ten to make up for that.

For example, if you measure a voltage that makes the trace rise 3.5 divisions, and your attenuator is set to 0.1 volts/div, multiply 3.5 × 0.1 × 10 to get 3.5 volts. That might sound overly complicated, but there's an easy way to do it: just move the decimal point one space to the right when looking at the attenuator. If it's set to 0.1 volts/div, remember it's really reading 1 volt/div. Then multiply that by what you see on the screen, and you're all set.

Some fancier scopes automate the 10X factor when used with their own brand of probes. The probes alert the scope to the scaling factor, and the input attenuator's markings are illuminated at the correct spot, so you don't have to do the arithmetic.

When you're interpreting the displayed signal, keep in mind that the precision to which you can measure things depends on the setting of the attenuator. If it's set to 1 volt/div (after accounting for the 10X probe factor, of course), you can visually estimate down to about 0.1 volt using the subdivision lines and the spaces between them. If it's set to 10 volts/div, you can estimate only down to about 1 volt, since the same graticule box now represents 10 volts instead of 1, so each subdivision represents 1 volt.

You may be wondering why probes divide the signal by 10, and why some have switches. To display a signal, the scope has to steal a tiny amount of it from the circuit you're testing. It's like a blood test: you have to take a little blood! The object of the probe's division is to present a very high *impedance* (essentially, resistance) to the circuit to avoid loading it down—that is, stealing enough current from it to alter its behavior and give you a false representation of its operation. Because the 10X probe needs to steal only a tenth of the signal's actual voltage, it has internal voltage-dividing resistors that give it an impedance of 10 MΩ, or 10 million ohms. That's very high compared to the resistances used in any common circuit of the sort you'll want to measure. Extremely little current passes through such a high resistance, so the circuit under test doesn't notice it.

If your probe offers 1X, that switch position removes the voltage divider, passing the signal directly to the scope and resulting in an input impedance of 1 MΩ. That's still pretty high, but it can affect some small-signal and high-frequency circuits. Usually, you'll keep your probe at 10X unless the signal you want to see is so small that you can't get enough vertical deflection on the screen even with the attenuator set to its most sensitive range. Now and then, you may scope a circuit that generates enough electrical noise to get into the probe through the air, like a radio signal. When you're at 10X, the impedance is so high that it takes very little induced signal to disturb your measurement. Switching to 1X may make the extraneous noise disappear, or at least get much smaller. This kind of thing happens mostly when probing CRT TV sets, LCD backlighting circuits and switching power supplies, all of which use high-voltage spikes capable of radiating a significant short-range radio signal.

Using 1X will let you see rather small signals as long as it doesn't load them down too much. Also, some scopes let you pull out the center knob on the attenuator to multiply the sensitivity of the selected range by a factor of 10. Doing so causes some signal degradation, so use this only when you really need it. You probably never will.

Ah, that center knob. It's called the *variable attenuator*. Normally, you keep it in the fully clockwise, calibrated position so the volts/div you select will match the graticule, allowing you to measure voltage values. Sometimes you want to do a relative measurement—that is, one whose absolute value doesn't matter, but you need to know if it's bigger or smaller than it was before, or its size relative to another signal. To make such measurements easy to read, it's very helpful to line up both ends of the signal with lines on the graticule. Turning the center knob counterclockwise gradually increases the attenuation, reducing the vertical size of the signal and letting you align its top and bottom with whatever you like. It's crucial to remember, though, that you can't take an actual voltage measurement this way; the vertical spread of the signal has no absolute meaning whenever the variable attenuator is engaged. To remind you, many scopes have a little "uncal" light near the attenuator so you'll know when the variable attenuation is on and the channel is uncalibrated.

Input Coupling (AC/DC/GND)   This determines how the signal is *coupled*, or transferred, into the vertical amplifiers, and it's one of the most important options on a scope. As you will see, the choice of coupling enables a neat trick for examining signal details and is not limited to being used in the obvious way, with DC for DC signals and AC for AC signals. You'll find yourself switching between the two settings quite often when exploring many types of signals.

GND   The GND setting simply grounds the input of the scope, permitting no voltage or signal from the probe to enter, and discharging the coupling capacitor used in the AC setting. (More about that shortly.) It does *not* ground the probe tip! You don't need to remove the probe from the circuit under test to switch to the GND setting. This selection is used to position the trace at a desired reference point on the screen, using the input channel's vertical position control, with no influence from incoming signals. Not all scopes have a GND setting. Some probes offer it on their 10X/1X switches. If you don't have one in either place, it's no big deal, because you can always touch the probe tip to the ground clip. Having a GND setting just makes getting a clean, straight line of 0 volts a bit more convenient.

DC   In the DC position, the signal is directly coupled, and whatever DC voltage is present will be plotted on the screen. If you want to measure the voltage of a power supply or the bias voltage on a transistor, use this position.

To measure a DC voltage, first remove the probe from the circuit and touch it to the ground clip, or switch to the GND setting, and turn the vertical position knob to wherever on the screen you want to call 0 volts. If you're measuring positive voltage, the bottommost graticule line is a good place to put the trace. If the voltage is negative, set the trace at the topmost line, because it will move down when the signal is applied. Then touch the probe to the point you want to measure, or switch back from GND to DC, and observe how many graticule divisions the trace rises or falls. Don't forget to multiply the attenuator's marking by ten to compensate for your 10X probe!

If you touch a voltage point with the probe and the trace disappears, you have probably driven it off the screen with a voltage bigger than can be handled by the range you've selected with the vertical attenuator control, so set that to higher voltages per division until you can see the trace. After switching the attenuator to a different range, perform the zero setting again before trying to estimate a measurement from the screen, because sometimes it drifts a little bit when you change the range.

AC   Many signals contain both DC and AC components. They have a DC *voltage offset* from 0 volts, but they're also not just a straight line; there's variation in the voltage level, representing information or noise. So how is a varying DC voltage AC? Isn't it one or the other?

As I mentioned in Chapter 5, voltage level and polarity are entirely relative. Any voltage can be positive with respect to one point and negative with respect to another.

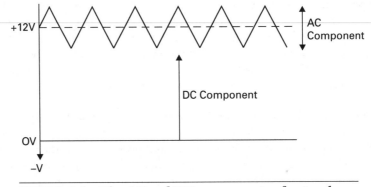

FIGURE 6-3   The AC and DC components of a signal

Take a look at Figure 6-3. That signal is above ground in the positive direction, with none of it going negative, so it's a DC signal with respect to ground, and its height above ground is its *DC offset* or *DC component*. The top of it, however, is wiggling up and down. Imagine if you could block the DC offset, and that dotted line became 0 volts. The signal would then be AC, with half above the line and half below.

It looks nice on paper, but you can't actually do that, right? Sure you can! Passing the signal through a capacitor will block the DC level, but, as the signal wiggles, those changes will get through, resulting in a true AC signal swinging above and below ground, with polarity going positive and negative. Hence, those wiggles are known as the *AC component* of the DC voltage. (And now you know a quick-and-dirty way to turn a positive voltage into a negative one! See, I wasn't kidding, there truly is no absolute polarity.)

When you select AC coupling, the scope inserts a *coupling capacitor* between the probe and the vertical amplifier, blocking any DC voltage from deflecting the beam. This changes everything! By blocking the DC component of a signal and passing only the AC component, you can examine that component in great detail. Let's see how.

Suppose you have a 12-volt power supply in an LCD monitor with erratic operation. The backlight doesn't like to turn on. When it finally does, sometimes it shuts itself off. You suspect the supply might not be putting out clean power. In other words, some noise could be riding on top of the voltage, perhaps caused by weak filter capacitors, confusing the microprocessor and turning it off. You fire up your scope, set it to DC coupling and check that 12-volt line, but it looks okay. Hmmm...there might be a little blurriness on the line, but it's hard to tell for sure, so you crank up the sensitivity of the vertical attenuator to take a closer look. Oops! The trace is now off the screen. You turn the vertical position control down to get it back, but now that control is as low as it will go. You can see the line, but you can't check for any small spikes or wobbles in it because it keeps going off the screen every time you up the sensitivity enough to examine the small stuff.

If only that same noise were riding on 0 volts, instead of 12, then the trace would stay put and you could fill the screen with even the tiniest changes. No problem. Switch to AC coupling to block the DC component of the signal and you can crank the attenuator for maximum sensitivity without budging the trace. It's a very powerful technique for examining signals with both DC and AC components. Many signals are of that form, and you'll use AC coupling quite often, regardless of whether the signal is really AC or not. In fact, a true AC signal (one that swings positive and negative with respect to ground, with no DC offset) will read exactly the same with either DC or AC coupling, so switching between the two is an easy way to see if an offset exists. If the trace doesn't shift vertically when you flip the coupling switch, there's no offset.

If the signal's changes are slow enough, the coupling capacitor charges up and eventually begins to block the slowly changing signal voltage until the other half of the cycle discharges it. This is called *low-frequency rolloff*; the cap acts as a high-pass filter, permitting high frequencies to pass through, while gradually rolling off (attenuating) lower ones, passing nothing when the frequency reaches zero. The effect will distort low-frequency signals, causing their flat areas to droop as the cap charges. See Figure 6-4.

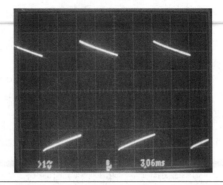

FIGURE 6-4   Tilt due to low-frequency rolloff

To see this in action, view about four cycles of the square wave from the cal terminal and switch between AC and DC coupling. (Adjust the vertical position if a DC offset moves part of the waveform off the screen when you change the coupling.) The flat tops and bottoms look tilted with AC coupling, losing amplitude and heading toward the middle of the waveform as the cap charges. As the signal frequency rises, the cap has less time to charge, so this effect fades away, for a truer representation of the signal. When you're viewing low-frequency signals with AC coupling, always keep in mind that long, sloping areas may in fact be flat, and the scope's coupling cap might be causing the slope. The easy way to verify the presence of rolloff is to switch to DC coupling (unless a DC offset drives the trace too far off the screen for the vertical position control's range to bring it back). If the sloping lines become flat, you know the cap was fooling you.

After using AC coupling, always ground the probe tip or momentarily switch the input coupler to GND to discharge the coupling capacitor before probing another point. Otherwise, whatever is stored on it will discharge into the next point you touch, confusing your reading and possibly damaging the circuit under test, although there's only a remote chance of that. When you ground the probe, you should see the trace jump for a fraction of a second and then return to where it was as the cap discharges.

AC coupling can be inconvenient when you're working with signals that change amplitude (vertical size) in an asymmetrical fashion. Analog video, for instance, is "clamped" to a fixed voltage, with its sync pulses not deviating from their position at the bottom as the video information at the top changes with the TV picture's content. Using AC coupling with such a signal will cause both ends of the waveform to bounce around as video content rises and falls, because the midpoint of the signal is constantly shifting. The effect is disconcerting and difficult to interpret. Video signals and others with similar asymmetry, like digital pulse streams, are best viewed with DC coupling, so that the unchanging end of the signal stays put.

The example was from a real case of an LCD monitor I fixed using the AC coupling on a DC signal technique. Sure enough, there were narrow, 1/2-volt spikes on the power supply line confusing the microprocessor and tripping the unit off. Once I switched to AC coupling and saw the spikes on the filter capacitor's positive terminal, I didn't even have to check the cap, because I knew a good one would have filtered them out. I just popped in a new cap. A quick check confirmed that the spikes were gone, and the noise on that line was now well under 100 mv. The micro was happy and so was I. The monitor worked great. Without the AC coupling trick, I'd never have known those spikes were there.

Vertical Position   This moves the trace up and down. Use it to set the zero reference point when taking DC measurements. Otherwise, set it wherever is best for examining the signal you're viewing. When using the scope in dual-trace mode, set the two input channels' vertical position controls to keep the waveforms separated. It's conventional to put channel one in the top half of the screen and channel two in the bottom, but you can place them wherever you want, even on top of each other.

Bandwidth Limit   This limits the frequency response of the vertical amplifiers. In 100-MHz scopes, the limit is usually 20 MHz. Switching in the limit removes noise bleeding in from external sources, particularly FM radio stations. With 100-MHz bandwidth, the scope can pick up the lower two-thirds of the FM band, resulting in a noisy-looking signal if you happen to live near an FM broadcast station. Some other radio services may get into your measurements too, if they're strong enough. Leave the bandwidth limit turned off unless noise problems are making your trace blurry.

Vertical Mode   On a dual-trace scope, this selects which channels you will view, and how. Most scopes offer these options:

- **Ch 1**   You will see only channel 1. Input from channel 2 can still be used to feed the trigger if you want, even though you can't see it.
- **Ch 2**   The same, but in reverse. You'll see only channel 2.
- **Add**   The voltages of the two channels will be added together and shown as one trace.

Adding two signals together is pretty pointless, so why is this here? One or more of your channels should have a button marked *invert* or *inv*. Pressing it makes the channel flip the waveform passing through it upside down, with positive voltages deflecting downward and negative ones upward. If you flip one channel upside down and then select the *add* mode, the signals will be *subtracted*, and that is very useful in certain circumstances.

Let's say you have an audio amplifier with some distortion. Or, perhaps, the high frequency response is poor. You want to find which stages are causing the problem. Feed some audio to the amplifier's input. Set both scope channels to AC coupling. Connect one channel to a stage's input and the other to the stage's output. If the amplifier hasn't already inverted the signal, do the invert-add trick. Then use the vertical attenuator of the channel with the bigger signal (usually the stage's output) to reduce its displayed amplitude until the signals cancel out and the resulting trace is as flat as possible. What's left is the difference between the two signals: the distortion.

Using this technique, you can easily see what a circuit is doing to a signal, and the results may be enlightening, leading you to a diagnosis. Most amplifier stages will show some difference, but it should be minor. When you see something significant, you've found the errant stage.

Chop   This puts the scope into dual-trace mode, displaying two signals at once. The two will appear to be independent, and you can set the vertical attenuation and position of each channel at will. The horizontal sweep speed set with the time/div control will apply to both channels, and the scope will trigger and start the sweep based on the timing of whichever of the two channels you choose for triggering. However, the scope is not truly displaying two simultaneous events; it just looks that way. In reality, it is rapidly switching between the two channels, with the beam bouncing between them, drawing a little bit of one channel and then a little bit of the other. It is chopping them up.

This method works very well, ensuring that the time relationship between the two signals is well preserved, since they are really being made by the same beam as it traverses the screen from left to right. It has a few serious limitations, however. If you crank the sweep speed way up, you can see the alternating segments of the two channels and the gaps between them. So, chop mode is not useful at high sweep speeds. Also, if the two signals are *harmonically unrelated* (their frequencies are not a simple ratio, so the cycles don't coincide), only the channel chosen for triggering will be visible; the other will be a blur.

Alternate or Alt   This is the other dual-trace mode, and it works rather differently. Instead of chopping the waveform, it draws an entire channel and then goes back and draws the other one, in two separate, alternating sweeps. Alt mode leaves no gaps in the traces, so it's suitable even for very fast sweep speeds. Plus, under certain circumstances, you can view harmonically unrelated signals, with separate triggering for each channel making them both look stable.

Alt mode has its own limitations, though. At slow sweep speeds, you'll see the two sweeps occur, one after the other, with the previous one fading away, resulting in a rather uncomfortable blinking or flashing effect. Also—and this one is much more serious—the timing relationship between the two signals may be disturbed, because of when the scope triggers and how long it takes to sweep the screen before it has a chance to draw the second waveform. In alt mode, the displayed alignment in time of one channel to the other *cannot be trusted*. Always choose chop mode when you need to compare or align the timing of two signals. In fact, use chop mode as much as possible, and select alt mode only when chopping interferes with the waveform or the two signals you're viewing are unrelated in time, so each one needs its own trigger.

X-Y   This mode stops the sweep and lets you drive the horizontal deflection from a signal input to channel 2. Be careful when pressing this button, because the stopped beam will create a *very* bright spot on the screen and can quickly burn the tube's phosphors. Turn the brightness all the way down before trying it, and then gradually turn it up until you see the spot.

At one time, X-Y mode was a useful way of detecting nonlinearity, measuring frequency and some other parameters. Today, it has little or no application, at least not for general service work. I have never needed to use X-Y mode even a single time.

Trigger Controls

To present a stable waveform, the scope must *trigger*, or begin each sweep, at the same point in each cycle of the incoming signal, so the beam will draw the same signal features on top of the last ones. Otherwise, all you'll see is a blur. Stable triggering is one of the most critical features on a scope, and it's important that you get good at using the trigger controls to achieve it.

Trigger Lock Light   This indicator tells you when the trigger is locked to a feature of the signal. When it's on, you should see a stable waveform. If not, some other control is improperly set, or the trigger may be locking to more than one spot in each cycle of the waveform.

Source   This selects which channel will be used to trigger the sweep, along with some other options:

- **Ch 1**   Channel 1's signal will feed the trigger. Use this mode for single-channel operation or for dual-channel work when you want channel 1's signal to control timing.
- **Ch 2**   Channel 2's signal will feed the trigger.
- **Alt**   In alt mode, each channel will feed the trigger, one after another. This completely invalidates the timing relationship between displayed signals, because you have no idea how much time has elapsed between when the first channel's sweep finished and when the second channel's sweep began. It's useful when you want to look at two signals whose periods are not related, and you want them both to display stably. Just remember that it's like having two separate scopes; no time relationship exists between the displayed signals.
- **Line**   This uses the 60-Hz AC line as a timing reference, generating 60 sweeps per second. It's useful when viewing signals at or very near that frequency whose own features make for difficult triggering. Now and then, it can be handy when troubleshooting line-operated gear, especially linear power supplies.
- **External or ext**   Many scopes have extra inputs that can be used as vertical channels and/or trigger inputs. External triggering is great for locking very complex signals the normal trigger can't get a grip on.

  For instance, when adjusting VCR tape paths while viewing the RF (radio-frequency) waveform from the video heads, there is no stable way to trigger at the start of each head's sweep across the tape, using the signal it produces. Instead, you must drive the scope trigger from another signal in the VCR that is synced to the headwheel rotation. The external trigger input provides a place to feed it in, keeping channel 2 (from which you could accomplish the same thing) free for viewing other signals.

Coupling   This is somewhat like the input coupling on the vertical amplifiers, but it offers a few more choices specific to triggering needs:

- **DC**   The trigger will lock to a specific signal voltage relative to ground. It's most useful with asymmetrical waveforms.
- **AC**   The trigger will lock to a voltage above or below the midpoint of the signal, regardless of its voltage relative to ground. It works just like the AC coupling option on the vertical amplifiers, placing a coupling capacitor in line to block the signal's DC component. You can use AC trigger coupling even when you use DC coupling for the vertical channel. Most of the time, you'll use this mode because it makes triggering easy as you look at various signals with different DC components.
- **HF reject**   This feeds the signal through a low-pass filter, smoothing out high-frequency noise or signal features that might confuse the trigger and cause it to trip where you don't want it to. If your waveform has high-frequency components causing jittery display, try this option. If you *want* to trigger on a high-frequency feature, however, selecting HF reject will prevent triggering. Using this setting affects only the trigger operation; the signal going to the vertical amplifier is not filtered.
- **LF reject**   This feeds the signal through a high-pass filter, rolling off low frequencies and preventing them from tripping the trigger. Use it to help trigger on high-frequency signal components when lower-frequency elements are causing triggering where you don't want it. Again, the filtering affects only the trigger.
- **TV-H**   This is a specialized trigger mode for use with analog TV signals. It comes from a time when much service work was on TV sets, and not all scopes offer it. It is optimized to help the scope trigger on the horizontal sync pulses in the TV signal.
- **TV-V**   This enables triggering on the vertical sync pulses in a TV signal.

Slope   This selects whether the trigger locks on signal features that are rising or falling. With many kinds of signals, like audio and oscillators, it doesn't matter. Normally, leave it on +. When you want to trigger on the falling edge of a waveform, switch it to –. To see the slope feature in action, connect the probe to the cal terminal, get a locked waveform and then switch between the slopes. Look at the leftmost edge of the screen to see what the waveform was doing when the sweep triggered.

Level   This sets the voltage level at which the trigger will trip. It has no calibration. Just turn it back and forth until the trigger locks. If the trigger stays locked through a wide swath of this control's range, that indicates a solid trigger lock, and you should see a very stable display. If you can get trigger lock only over a very narrow range of the level control, the lock is not great, and you can expect the waveform to jump around if the signal level or shape changes even a little bit. To get a better grip on the signal, try the various coupling options, especially HF reject and LF reject, rotating the level control back and forth for each one.

Holdoff   This keeps the trigger held off, or unable to trip, for an adjustable period of time after its last trigger event. It's used on complex signals with irregularly spaced features of similar voltages, to avoid having more than one in each cycle trip the trigger and blur the displayed waveform. If you can't get a waveform to stabilize any other way, try turning this control. Otherwise, keep it at the "normal" position, which reduces holdoff to a minimum. Forgetting and leaving it on can make triggering confusingly difficult, as the holdoff will make the trigger miss consecutive cycles if it's set for too long a period, resulting in a jumpy mess.

Horizontal Controls

The horizontal axis represents time, and the scope's drawing of the waveform from left to right is called the *sweep*. The settings controlling it determine what you'll see, even more than do those for the vertical parameters.

Horizontal Position   This positions the trace left and right. Set it to fill the screen, with the left edge of the trace just off the left side. Now and then you may wish to line up a signal feature with the graticule to make a rough measurement of period or frequency, and you can use this control to do so.

Sweep Mode   This control offers three options: auto, normal and single.

- **Auto**   The trigger will lock to the incoming signal, and sweep will begin. When there's no signal, or the trigger isn't locking on it for some reason, the sweep will go into free-run mode, triggering itself continuously so you can see a flat line, or, in the case of trigger unlock, a blur. This is the mode you will use most of the time.
- **Normal**   Sweep will begin when the trigger locks on a signal but will stop when the signal ceases. This can be handy for observing when rapid interruptions occur in intermittent signals, because the trace will blink at the moment the signal disappears. It's a little disconcerting sometimes, though, because if the screen goes blank, you don't know why. Maybe the trace is off the screen, maybe the brightness is too low, or maybe the trigger isn't locked.
- **Single**   This is for single-sweep mode, in which the trigger will initiate one sweep and then halt until you press the reset button (which may be the *single* button itself). It helps you determine when a signal has occurred, because you'll see the flash of one sweep go by. Look for a little indicator light near the *single* button labeled Ready or Armed. When it's lit, the sweep can be fired one time. After that firing, the light will go out, and you must press the reset button to rearm the sweep. You won't use this mode very often.

Time/Div   The outer ring of this large control sets the *timebase*, or sweep speed at which the beam will travel across the screen from left to right. It is calibrated by time in seconds, milliseconds (ms, or thousandths of a second) and microseconds (μs, or millionths of a second). The calibration number refers to how long the beam will take to traverse one division, or graticule box.

For a signal of a given frequency, the faster you set the sweep, the more horizontally spread the display will be, and the fewer cycles of the signal you will see at one time. Often, you will want to set it as fast as possible to see the most detail, but not always. In some instances, the aggregate effect of many displayed signal cycles can be more revealing than is a singular signal feature. If you have low-frequency variation, such as AC line hum, on a fairly fast signal, you can't see the hum if the sweep is set fast enough to see individual cycles of the signal. But when you turn the sweep speed low enough to view a 60-cycle event, you will see the hum clearly, even though the signal itself will be too crammed together to be resolvable.

To see the effects of various sweep rates, look at the calibrator's square wave and click the time/div knob through its ranges. On most scopes, the number of displayed waves will grow or shrink with the sweep rate. If it doesn't, your scope is changing the frequency of the calibrator to match the timebase. Some of them do that.

Variable Time   The center knob uncalibrates the timebase, slowing it down as you turn the knob counterclockwise. This is the horizontal equivalent of the vertical channels' variable attenuators, and it can be useful for lining events up with the graticule during relative time measurements of two signals. Normally, you'll leave it in the fully clockwise position. As with the variable attenuators, it has an "uncal" light to remind you that the timebase no longer matches the graticule.

Pull X10   On most scopes, pulling out the variable time knob speeds up the sweep rate by a factor of ten. The sweep does remain calibrated in this mode. It's a quick-and-dirty way to spread out a signal, and it also lets you get to the very fastest sweep rate by turning the time/div knob all the way up and then pulling this one out. Normally, keep this knob pushed in, and pull it only when you really need it. Don't forget to push it in again, or your time measurements will be off by a factor of ten.

Delayed Sweep Controls

Delayed sweep is an advanced scope function you won't need for basic repair work, but it's worth learning for those more complex situations, like servicing camcorder motor control servos, in which it's essential. My apologies for any neck injuries caused by making your head spin while reading this section! Once you actually play with delayed sweep a few times, you'll discover it's really not difficult, and it's quite nifty.

With delayed sweep, your scope becomes a magnifying glass, allowing you to zoom in on any signal feature, even though it's not the one on which you're triggering the main sweep. Why do this? It's very powerful, providing a level of signal detail you couldn't otherwise examine.

Let's say you have a sine wave from an oscillator, but it doesn't look quite right. A spike or something is distorting its shape at a particular spot. It's hard to tell what it is, but you can see a thickening of the trace at that point. You want to get up close and personal with that spot so you can really see the details of the distortion and determine what's causing it. You scope the sine wave and crank up the sweep rate, but when you get it going fast enough to spread out the mystery spot, it has already gone off the right side of the screen. You turn the sweep rate down a little bit, but now the signal is too crammed together to permit a good look at the spot.

Cue superhero music. This is a job for...Delayed Sweep, Slayer of Stubborn Signals, Vanquisher of Villainous Voltages! In this mode, the scope triggers on the waveform as usual, as set by the A trigger. It begins sweeping at the rate selected with the main time/div knob. After a period of time you set with the *delay time multiplier* knob, the B timebase takes over and the beam finishes the sweep at the speed set by that timebase's knob, the one inside the main sweep's knob.

The result is a compound view of the signal, with a lower sweep rate for the events leading up to your spot of interest, followed by a stretched-out, detailed view of the spot! Even cooler, you can rotate the delay time multiplier knob and scan through the entire signal, examining any part of it. It's practically a CAT scan for circuitry, and you don't even need a litter box. Let's look at the controls involved with setting up the delayed sweep mode.

Delay Time or B Timebase   This control, located inside the main time/div knob, sets the speed of the B sweep, which stretches out the waveform for close examination. Think of it like a zoom lens on a camera: the faster you set it, the more you're zooming in for a closer look at a smaller area.

On some scopes, notably those made by Tektronix, the same knob is used for both the A and B sweeps. To engage the B sweep, you pull out the knob, mechanically separating the two timebase controls. In this position, the outer ring, which sets the A sweep rate, will not move when you turn the knob clockwise to speed up the B sweep.

Horizontal Display Mode   This selects which timebase (sweep generator) will drive the beam across the screen, and is how you choose between normal and delayed sweep modes.

- **A**   The main timebase, which is set by the big time/div control, will control the sweep. Delayed sweep mode will not be engaged.
- **B**   The secondary timebase, set by the smaller control inside the time/div knob, will control the sweep. This may be set equal to or faster than the main timebase, but not slower.
- **Alt**   Both timebases will be displayed, one on top of the other. The A sweep's display will be highlighted over the area that the B sweep is stretching out. The faster you make the B sweep rate, the narrower the highlight will get, and the more stretched the displayed B sweep will be. See Figure 6-5. Look for a knob called *trace separation*. It lets you position the B sweep's display above or below that of the A sweep, so you can see them better. It has no calibration or meaning in voltage measurement terms; it's just a convenience.

  If your scope has two brightness controls, you may need to increase the brightness of the B sweep to keep it visible at high sweep rates. The faster the beam sweeps, the dimmer it will appear.
- **A intens B**   This stands for "A intensified by B" and shows you the A sweep and the highlight of the portion of the signal the B sweep will cover, just as is shown in Alt mode. It's useful for zeroing in on the spot you wish to examine without cluttering up the screen with the B trace, before you switch to a mode that displays B.

  In both Alt and A intens B modes, the highlighting of the A sweep may not be visible if you have the A brightness control set too high. Try turning it down for a more prominent highlight.

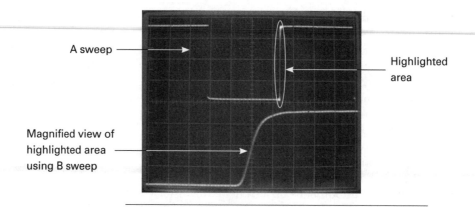

A sweep

Highlighted area

Magnified view of highlighted area using B sweep

FIGURE 6-5 Alt mode delayed sweep display of the rising edge of the scope's calibrator square wave. The A trace is on the top, with the B trace below it. Notice the highlighting of the portion of the waveform being stretched out by the B trace.

- **B** This shows only the B trace, at the sweep rate selected by the B or delay time setting (the knob inside the big, outer time/div one). It turns off the A trace, but the settings which result in the delayed sweep B view remain in effect, including triggering, A sweep rate and delay time. Once you've zoomed in on your spot of interest, you can switch to this setting to see only the detailed area, without the main waveform from which it is derived. In this mode, the trace separation control doesn't do anything.
- **Start after delay/triggered** This determines what happens after the A sweep has reached the beginning of the intensified area where it will switch to B. In *start after delay* mode, the B sweep begins as soon as the A sweep hits that point. This lets you scan continuously through the signal with the delay time multiplier control, and is my preferred way of using delayed sweep. It has one drawback, though. Any jitter (instability) in the triggering of the A sweep will get magnified by the faster sweep rate of the B sweep, resulting in the stretched waveform's wobbling back and forth. Sometimes the wobble is bad enough to make examining the signal difficult.

To remove the wobble, select the *start triggered* mode. Then, when the A sweep reaches the intensified portion, the B sweep will wait for a trigger before beginning. This is what those B trigger controls are for. What's great about having a separate trigger for this function is that you don't have to use the same trigger settings for the B sweep that you selected for the main trigger. So, you can treat the detailed area as a separate signal, setting the trigger for best operation on its particular features.

The downside to this mode is that you cannot scroll smoothly through a signal, because the B sweep waits for a trigger. You can see only features on which it has triggered. Keep your scope in *start after delay* mode unless you run into wobble problems or are examining fine details in a complex signal, like an analog video signal, with many subparts.

Delay Time Multiplier   This sets how many horizontal divisions the delay will wait before starting the B sweep (or looking for a new trigger for it, in the *start triggered* mode), and is referenced to the A sweep rate. For example, if the A sweep is set to 0.1 ms and the delay time multiplier knob is at 4.5, it'll wait 0.45 ms before the highlighted area begins. For most work, you can ignore the numbers and just keep turning the knob to scan through the signal, using the highlight to pick your expansion area. Most scopes provide a vernier (gear reduction) knob for fine control.

B Ends A   This turns off the A sweep after the B sweep starts. On some scopes, it presents both sweeps in one line, switching from A to B when B begins. On others, the two sweeps are still on separate lines, and A just disappears after the point where B takes over.

Try It!

How are those neck muscles doing? Ready to try some of this? Let's use delayed sweep to get a close-up look at the rising edge of your scope's calibrator square wave.

Connect the probe to the cal terminal and set up the scope for normal, undelayed operation by selecting the A horizontal display mode. Adjust the vertical, trigger and time/div controls to see a couple of cycles on the screen, and put them on the top half.

Disengage B ends A mode if it is currently turned on. Make sure start after delay is engaged, not start triggered. Select the A intens B mode and move the delay time multiplier knob to center the highlight over the leading (rising) edge of the square wave, with the start of the highlight just before (to the left of) the edge. Set the brightness controls as desired so you can see the highlight clearly. The A brightness control will adjust the bulk of the waveform, with the B brightness setting the highlight's intensity.

If more than one cycle of the waveform is on the screen, any of the leading edges will do. It's best to keep the A sweep as fast as possible, though, without losing the feature you want to examine off the right side, so try not to have more than one or two cycles visible.

Adjust the delay time (B sweep rate) so that the highlight is just a little wider than the edge, covering a smidge of the waveform before and after it. Note that the faster you make the B sweep, the narrower the highlight gets, meaning you will be zooming in on a smaller area, enlarging it proportionally more.

Now switch to alt mode. You should see the zoomed-in rising edge of the waveform superimposed on the original wave. Use the trace separator knob to move it down so you can see it clearly. Increase the B sweep rate one step at a time. As you crank it up, the magnified edge will move off the right side of the screen, and you'll have to turn the delay time multiplier knob clockwise just a tad to bring it back. If the magnified waveform gets too dim, crank up the B brightness.

When you get the B sweep going pretty fast, you can clearly see the edge's slope, and you may also notice some wobbles or other minuscule features at its bottom and top, details you can't see at all without the magic of delayed sweep! For fun, scan through the waveform with the delay time multiplier knob, and take a look at the falling edge too. Pretty awesome, isn't it? Enjoy! Just don't forget to turn the A brightness down when you switch back to normal, undelayed operation, if you turned it up.

Cursor Controls

If your scope offers numerical calculation, it will have cursor controls that let you specify the parts of the waveform you wish to measure. The results of the measurements will be shown as numbers on the screen, along with the waveforms.

The layout of controls can vary quite a bit in this department, but the principles are pretty universal. For amplitude measurements, you move two horizontal cursors up and down to read the voltage difference between them. For time measurements, you move two vertical cursors left and right to measure the time difference between them, or to calculate approximate frequency. See Figure 6-6.

On some scopes, you can lock one cursor to the other after you've set them, so you can move one to the start of a waveform feature and the other will follow, letting you see how the signal aligns against the second one.

Always keep in mind that these measurements provide nowhere near the accuracy or precision of those you'll get from your DMM or frequency counter! Still, you can't measure the voltage or frequency of items within signals with anything but a scope.

Digital Differences

Operating a digital scope isn't that much different from using an analog, but there are some items to keep in mind.

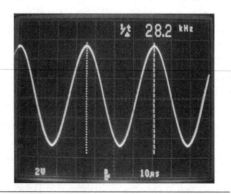

FIGURE 6-6 Cursors and frequency (1/time) measurement on a Tektronix 2445 analog scope

First, many of the controls on an analog instrument are replaced by menus on its digital counterpart. This approach unclutters the front panel, but it's slower and more awkward to have to step through nested menus than it is to reach for a knob. As long as you keep the basic functions of vertical, horizontal and triggering in mind, though, you should have no trouble remembering where to find the options you need.

All the screen controls are gone. You don't need astigmatism, focus, illumination and separate A and B brightness, because the display is an LCD screen and it is not being swept at varying rates by a beam. You'll find a main brightness or contrast adjustment in a menu. Once set, it won't require any changes for different modes or sweep speeds.

The display will show various operating parameters like trigger status, volts/div and time/div, so you don't have to take your eyes off the screen and interpret a bunch of controls.

Digitals are generally more accurate, especially in the horizontal (time) domain. They also tend to have more stable triggering and very little drift.

Delayed sweep may be handled a bit differently. On the Tektronix TDS-220, for instance, the equivalent of the analog scope's highlight is called the Window Zone, and you set the delay time multiplier and window width using the horizontal position and time knobs after selecting the mode from a menu. Instead of a highlight, you get a set of long, vertical cursors. Then you select Window to see the magnified waveform. The principles are the same, of course, but there's no equivalent to the analog instrument's alt mode, in which both the undelayed and delayed sweeps are shown simultaneously.

More than likely, the digital scope will include various acquisition modes, so you can grab waveforms, store them and display them along with live signals. It'll probably also have measurement options for frequency, period, peak-to-peak voltage, RMS voltage and so on. Especially nice is the auto setup function, which sets the vertical, horizontal and trigger for proper display of a cycle or two of most waveforms, without your having to twist a single knob. Hook up the probe, hit the *auto* button and there's your signal. It's the oscilloscope equivalent of autoranging on a DMM, and it saves you a lot of time and effort.

Reading the screen on a digital scope requires more interpretation. The limitations imposed by the digital sampling process, the finite resolution of the dot-matrix display and the slower-than-real-time screen updating have to be kept in mind at all times. For one thing, curved areas can have jagged edges, and it's important to remember that they are not really in the signal. Also, lines may appear thicker and noisier than they really are, due to digital sampling noise. Aliasing of the signal against the sampling rate and also against the screen resolution can seriously misrepresent waveforms under certain circumstances, as discussed in Chapter 2. Finally, the slow updating means some details get missed. All scopes, analog or digital, show you snapshots, but with digitals there's much more time between snaps.

When the input signal disappears, many digital scopes freeze the waveform on the screen for a moment before the auto sweep kicks in and replaces it with a flat line. This makes it hard to know exactly when signals stop. If you're wiggling a board while watching for an intermittent, the time lag can hinder your efforts to locate the source of the dropout.

Overall, a digital instrument offers more stability, options and conveniences, but an analog scope gives you a truer representation. As with just about everything, digital is the future, like it or not, and analog scopes will probably disappear from the marketplace in the next few years. If you snagged a good one, hang onto it! If you chose a digital instrument, just keep these caveats in mind and you'll be fine.

Soldering Iron

Soldering is probably the most frequent task you'll perform in repair work. It's also one of the easiest ways to do damage. Competent soldering technique is essential, so let's look at how to do it.

Never solder with power applied to the board! The potential for causing calamity is tremendous. First, you may create a path from the joint, through your iron to ground via the house wiring, resulting in unwanted current. Second, it's very easy for the iron's tip to slip off the joint and touch other items nearby. Make sure power is truly disconnected, remembering that many products don't actually remove all power with the on/off switch. Unplug the item or remove the batteries to be sure.

A good solder joint is a molecular bond, not just a slapping of some molten metal on the surface. The solder actually flows into the metal of the component's leads and the copper circuit board traces. When it doesn't, the result is called a *cold solder joint*, and it will fail fairly quickly, developing resistance or, in some cases, completely stopping the passage of current.

To get a good joint, first *tin* the iron's tip. Warm up the iron to its full temperature and then feed a little bit of solder onto the tip. It should melt readily; if not, the tip isn't hot enough. Coat the tip with solder—don't overdo it—and then wipe the tip on the moistened sponge in the iron's base. If you have no sponge, you can use a damp (not dripping wet) paper towel, but strictly avoid wiping the tip on anything plastic. Melted plastic contaminates the tip badly and is tough to remove.

Once the tip is nice and shiny, put another small drop of solder on it. Press the tip onto the work to be soldered, being sure it makes contact with both the circuit board's pad and the component's lead (or contact point, in the case of surface-mounted, leadless parts). Then feed some solder into the space where the lead and the pad meet, until you have enough melted solder to cover both without creating a big blob. See Figure 6-7. To get a good idea of how much solder to use, look at the other pads. See Figure 6-8.

As the solder feeds, it should flow into the metal. Check around the edges for smooth integration into the joint. If you see a ring of brown rosin, gently scrape it away with an X-Acto knife or very small screwdriver so you can get a good look. Also check for proper flow around the component lead. Sometimes the flow is fine to the board's pad, but the solder is pooled around the lead without having bonded to it because the iron's tip didn't make good enough contact to get it adequately hot, or the lead had a coating of oxidation that blocked the necessary chemical bonding. In fact, that style of cold solder joint is a major cause of factory defects resulting in warranty claims. At least it was, back when most components had leads. It doesn't occur nearly as often with leadless, surface-mount parts. Frequently, though, the problem is where the solder meets the pad. See Figure 6-9.

FIGURE 6-7   Proper soldering technique

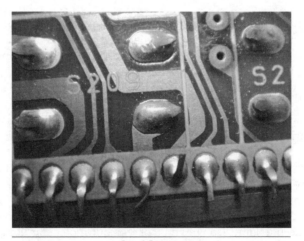

FIGURE 6-8   Good solder joints

If the joint looks like a bead sitting on top of the pad, you have not created a molecular bond and will need to reapply the iron. To get enough heat, the wattage of the iron has to be appropriate to the size of the joint. Also, you have to apply some pressure to the tip for effective heat transfer; a very light touch won't do it. Don't press really hard, though, as it won't improve transfer and could cause damage.

If the iron's tip is contaminated, heat transfer will be limited. It should look shiny. Especially if it has come in contact with plastic, it could have a coating blocking the heat. Although plastic contamination is most easily removed by scraping the tip when the iron is cold, tinning and then wiping a hot tip may cut through the coating.

FIGURE 6-9 A cold solder joint. Note how the edge fails to flow into the surrounding pad.

Contamination can occur on the leads of replacement components, too, especially with parts that have sat around for years in your parts drawers. If the leads look dull, apply some fine sandpaper or scrape them clean before attempting to solder them. They should be shiny for good solder flow.

Once you have a nice, properly flowed joint, remove the supply of solder and then the iron, in that order. If you do have to reflow the joint, add a small amount of new solder so you'll have fresh rosin on the joint to help facilitate bonding.

Soldering leadless, surface-mount components is tricky, mostly because they are so small that it's hard to keep them in place while applying the iron. Make sure the board's pads are completely flat, with no solder blobs on them, and then put the part in place. Hold it down with a small screwdriver placed in the middle of the component while you solder one end. Unless you're anatomically quite unusual, you won't have an extra hand to feed solder to the joint, so put enough solder on the tip to make a crude joint. Don't even worry about molecular bonding. Just tack that side down, even if it's with a bad joint. Then let go of the component and properly solder the other side, taking care to make a good joint. Finally, go back to the first side and do it right. You might have to wick off your first attempt before trying again. Be extra careful not to heat the part too much, or the good side will come unsoldered; those tiny parts conduct heat much faster than do larger components with leads. Also, a lot of heat can delaminate and destroy the component's solderable platings. What works best is adequate heat applied quickly. Get on and off the part with minimum delay.

 Before trying to solder tiny, surface-mount parts in a device you're trying to repair, practice on a scrap board. Experience really helps in developing successful soldering technique with these minuscule components.

FIGURE 6-10 Stranded wires twisted together

After soldering, the board will be left with a coating of rosin on and around the new joint. Some techs leave it on, but it's not a good idea because it can absorb moisture over time. Loosen it by gently scraping with the tip of an X-Acto knife or a small screwdriver. Wipe up what's left with a swab wet with contact cleaner.

To join wires, first twist them together for a solid mechanical connection. If the wires are stranded, try to separate the strands a bit and intertwine them when twisting the wires together. See Figure 6-10. If you're using heat-shrink tubing, keep it far from the soldering work or it'll shrink before you get a chance to slip it over the joint. And don't forget to slide the tubing onto one of the wires before entwining and soldering them! It really helps you avoid the expletives from having to cut the wires and start over.

Desoldering Tools

You'll use desoldering tools almost as often as your soldering iron. The two basic types are wick and suckers.

Wick

For small work, wick is the best choice. It's easy to control, doesn't splatter solder all over the area and doesn't run the risk of generating a static charge capable of damaging sensitive components. Its only real drawbacks are that it can't pick up a lot of solder at once, it's a tad expensive and it's not reusable.

To wick the solder off a joint, place the wick on the joint and heat it by pressing the iron to the other side. Applying a little pressure helps. In this case, *don't* put that extra drop of solder on the tip first or it'll flow right into the wick, wasting some of the braid's capacity to soak up the joint's solder.

When the wick saturates with solder, pull it and the iron away at the same time. If you remove the iron first, the wick will remain soldered to the joint. If desoldering is incomplete, clip off the used wick and try again.

When desoldering components with leads, it can help to form the end of the wick into a little curve and press it against the board so that, when heated, it'll push into the hole in which the lead sits and soak up the solder stuck inside. Be extra careful to keep the wick up to temperature until you pull it away, or you could lift the copper off the board, creating a significant problem.

Sometimes you are left with a film of solder the wick refuses to soak up. If that occurs, try resoldering the joint with a minimum of solder, just enough to wet it down a little. Then wick it all up. The fresh rosin of a new joint can help the wick do its job, carrying the old solder to the wick with it.

Suckers

Solder suckers come in several varieties. The most common are bulbs, bulbs mounted on irons, and spring-loaded.

Bulbs work well when there isn't a lot of solder to remove; they tend to choke on big blobs of it. To use a bulb, squeeze the air out of it, heat up the joint with your iron, position the bulb with its nylon tube directly over the molten solder, get the tube into the solder and relax your grip on the bulb. Although the end of the bulb is plastic, it won't contaminate your iron's tip because the plastic used is a high-temperature variety that doesn't melt at normal soldering temperatures.

After a few uses, the tube may clog with solder. Just push it inside with a screwdriver, being careful not to pierce the bulb. If it's so clogged that you can't budge the solder, pull the tube out and expel the plug from the other end. Eventually, the bulb will fill up and you'll have to remove the tube to empty it anyway.

When you have a large joint with lots of solder, a spring-loaded sucker is the only thing short of a professional vacuum-driven desoldering station that will get most or all of the solder in one pass. Cock the spring and then use it like a bulb.

The fast snap of a spring-loaded sucker can generate a static charge reputed to be capable of damaging sensitive parts, especially MOSFET transistors and integrated circuit chips of the CMOS variety. To be on the safe side, don't use one on those kinds of parts.

Rework Stations

Solder removal gets trickier as parts get smaller. Some of today's surface-mount parts, which have no leads poking through holes in the board, are getting so small that traditional soldering and desoldering tools are inadequate for working on them. Surface-mounted integrated circuits (IC chips), especially, may have dozens of leads so close together that manual soldering is impossible. To cope with the problem, advanced repair centers use rework stations. These systems have specialized tips made to fit various IC form factors, and they can resolder as well as desolder. Alas, rework stations are quite expensive and out of reach of most home-based repairers.

Power Supply

When powering a device from your bench power supply, you need to consider several factors for successful operation, and to avoid causing damage to the product.

Connection

Many battery-powered items also have AC adapter jacks offering a convenient place to connect your supply. The usual style of connector is the *coaxial plug*. See Figure 6-11. These plugs come in many sizes, and both inside and outside diameters vary. If you can find one in your parts bins that fits, perhaps from an old AC adapter or car adapter cord, you're in luck! Sometimes you can use a plug with a slightly different diameter, but don't force things if it isn't a good fit. You may find one that seems to fit but doesn't work because the inner diameter is too large, so the jack's center pin won't contact the inner ring of the plug.

The polarity of the plug is paramount! Don't get this backward or you will almost certainly do severe damage to the product as soon as you hit the supply's power switch. Coaxial plug polarity is usually printed on the device somewhere near the jack, and it will be on the AC adapter as well. Most modern products put positive on the tip and negative on the outer sleeve, but not all. Always check for the polarity diagram. It should look like one of these:

If there's no adapter jack, or you choose not to use it, you can connect clip leads to the battery terminals of most devices. This should work fine with anything using standard, off-the-shelf cells like AAs. The convention is to use a red lead for positive and a black one for negative, and I strongly suggest you do so to avoid any possible polarity confusion, which could be disastrous.

In a typical case, you open the battery compartment and find a bunch of springs and contacts, one set for each cell. Most of them link one cell's positive terminal to the next one's negative, forming a series string. One spring (the negative terminal) and one positive terminal (usually a flat plate or wire) feed the circuitry from each end of the string. Which are the two you need?

Sometimes the positions of the terminals offer mechanical clues. If you see two connected directly to the board, or if they're placed such that they could be, those are probably the right ones. Also, if one set of terminals is on a flip-out or removable door, that pair is *not* what you're looking for. If you find no such hints, use your DMM's continuity feature to determine which terminals are connected to adjacent ones. Whatever's left should be the two magic terminals.

If you can reach the terminals with probes while the batteries are installed, pull the cells out, use your DMM to measure each one and then reinstall them. Add up the voltages to get the total series voltage, and then look for it between terminals.

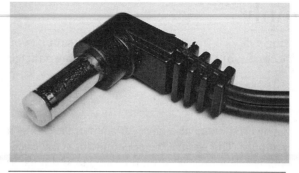

FIGURE 6-11 Coaxial power plug

Make sure the product is turned off, so it won't pull the voltage down. Only the two correct terminals will show your calculated voltage; any other combination will be significantly lower.

If the device uses a square 9-volt battery, the larger, petal-like terminal of the unit's snap-on connector is positive. Nine-volt gadgets are pretty rare these days, but you will encounter this battery style if you work on old transistor radios or tape recorders. Some digital answering machines and clock radios still use 9-volt batteries for memory backup.

In products with proprietary lithium-ion batteries, it's often possible to provide power through the device's terminals, but not always. Some of them, especially camcorders and laptops, use "smart" batteries containing their own microprocessors, and the device won't recognize power applied to the terminals without the data those micros provide.

Many smaller items, like digital cameras, may have three terminals. Two are for power, of course, and the third one is for a temperature sensor to prevent overheating during the charge cycle. Usually, these devices can be powered from a power supply, with the third terminal left unconnected.

To determine which terminal does what, look at the battery. You probably won't find any polarity markings on the product, but they are nearly always printed on the battery, and you can place it in the orientation required for insertion and see which terminals line up with the ones in the unit. You want the + and – terminals, of course. The other one may be marked "C" or have no marking at all.

If the battery is also unmarked, try measuring its voltage with your DMM. Most commonly, the voltage output is from the two outer terminals, with the sensor terminal between them. Once you find the right terminals, you'll also know the polarity. This assumes, of course, that the battery has at least a little charge on it; there's no way to read a dead battery, and don't even *think* of trying to apply a charge from your power supply without knowing the polarity. Lithium-ion batteries are nasty when they burst. You can get hurt. Even when you do know the polarity, putting too much current through lithiums too fast can make them go boom.

Voltage

Set the supply's voltage before connecting it to the device, and try to get it pretty close to what the unit expects. Most products list their battery voltages either on the back of the unit or on the battery, in the case of proprietary cells. In AA- or AAA-driven units, just multiply the number of cells by 1.5. Remember, though, that NiMH (nickel-metal hydride) rechargeable cells are only 1.2 volts each. If you run into the rare item that is made for use *only* with rechargeables, multiply by 1.2 instead. Some digital cameras and MP3 players fall into that category; they will not function with alkalines or other throw-away cells, and the higher voltage may damage them.

Most hobby-grade power supplies have analog meters, and they can be off by quite a bit. To keep things more accurate, use your DMM to set the voltage. Don't worry about millivolts; just stay within half a volt or so and you should be fine. Even lithium-ion batteries start out a little above their rated voltage when fully charged, with the voltage dropping as the charge is drained. The curve is a lot flatter than with other battery technologies, though, which is why it's a good idea to match the rated voltage the best you can.

Once you've set the voltage, turn the supply back off and connect the leads, double-checking the polarity. Then, hit the switch and pray. No smoke? Great! You're in business.

Current

The current drawn by a device will vary, depending on what the unit is doing. Especially with any product employing moving parts like a hard drive platter or laser optical head, current demand goes way up during mechanical motion, dropping again when movement ceases.

As long as your supply has sufficient current capacity, it doesn't matter. If, however, the supply has enough for some modes of a device's operation but not others, the results can be unpredictable.

This issue crops up during service of camcorders and hard drive–based MP3 players. The drive spins up or the tape loads, and suddenly the device shuts down or its micro gets scrambled due to the lowered voltage from an overloaded supply. If your supply has a current meter, keep an eye on it to be sure you're never pulling the supply's maximum current. If the meter does show maximum, the product's demand is probably exceeding what the supply can offer, and the voltage is dropping.

Transistor Tester

Using a transistor tester requires taking the transistor out of the circuit. Doing so ranges from easy, with small-signal transistors, to a hassle, with power transistors mounted on heatsinks. Transistors have three leads (see Chapter 7), so you'll have to disconnect at least two of them, though it's usually easier just to desolder all three and pull the part off the board. Big power transistors with metal cases employ the case as one terminal, usually the collector. With those, you may find it more convenient to leave the part on the board and disconnect the other two terminals, especially if they're connected with wires, rather than directly to circuit board traces.

There are several basic types of transistors, and testing procedures vary. With some transistor testers, you need to know which terminal is which, while others will try out all the combinations automatically and recognize when the correct configuration has been found.

Fancy transistor checkers can evaluate a transistor's characteristics in actual operation by using the transistor as part of an oscillator built into the checker. They can show you the part's gain and leakage. For most service work, such sophisticated measurement is unnecessary. Usually, you just want to know if the part is open or shorted. A transistor can be "leaky," passing reverse current or allowing flow between terminals when it shouldn't occur, but it doesn't happen often.

To test a transistor, connect its three leads as specified in your tester's instructions, and read whatever info it gives you. There are too many types of testers to detail their operation here.

Some DMMs include transistor test functions. If yours has a little round socket marked E, B and C, you have a transistor tester!

Capacitance Meter

Checking capacitors requires disconnecting at least one of their leads, because other circuit elements will distort the reading. Be absolutely sure to discharge the capacitor before testing it, especially with electrolytic caps, which can store a lot of energy capable of trashing your tester.

Turn the meter on and connect the capacitor to the input terminals. Some meters have special terminals into which you can press the cap's leads. You can use those or the normal clip leads, whichever is more convenient. If the capacitor is polarized, connect it the right way around, + to + and – to –! See Chapter 7 for polarity marking styles.

If your meter is autoranging, it'll step through its ranges and read the cap's value. If it isn't, begin at the most sensitive range, the one that reads *pf* (picofarads, or trillionths of a farad) and work your way up until the "out of range" indicator goes away and you get a valid reading. When you're finished, unhook the capacitor and discharge it. Very little energy is put into the component to test it, so you can short across its terminals without worry.

The meter will show you the value of the capacitor in fractions of a farad. Some types of capacitors, especially electrolytics, have fairly wide tolerances, or acceptable deviations from their printed values. Typically, an electrolytic can be off by 20 percent of its stated value even when new. Manufacturers deliberately err on the high side to ensure that filtering will be adequate when the caps are used in voltage smoothing applications, as many are. If the cap reads a little high, don't worry about it. If it reads a little low, that may be okay too. If it reads more than 20 percent low, suspect a bad cap.

And if you can't get rid of the out-of-range indicator on any scale, the cap is probably shorted or very leaky. Open caps will read as extremely low capacitance.

Only very specialized capacitance meters can read an electrolytic cap's equivalent series resistance (ESR). This parameter goes up as a cap ages, eventually interfering with proper operation and rendering the cap useless. A cap can still be bad even though it looks fine on a normal cap meter. But if it reads significantly below its intended value, or shows a short or an open, it is bad.

Signal Generator

A signal generator is used to replace a suspected bad or missing signal temporarily so you can see what its insertion will do to a circuit's behavior. Inserting a signal is a very handy technique when working on audio circuitry. It can also help you check clock oscillator function in digital gear or sub for a missing oscillator in radio equipment.

For audio testing, it's best to use a sine wave somewhere in the lower middle of the audio spectrum, at around, say, 1 kHz. Using a sine wave prevents the generation of *harmonics* that could damage the amplifier under test, the speakers or your ears.

For clock oscillator substitution, set the generator to the same frequency as the missing oscillator (it should be marked on its crystal), and use a square wave. Set the peak-to-peak voltage of the generator just below whatever voltage runs the chip normally doing the oscillating. Don't exceed it, or you could "latch" and damage the chip.

For radio oscillator substitution, use a sine wave at the frequency of the missing oscillator. It's probably best to feed the signal from the generator through a capacitor of around 0.01 μf to avoid loading down the radio's circuits. Set the peak-to-peak oscillator voltage to something less than the power supply feeding the radio's stage. Usually some fraction of a volt is plenty in this kind of experiment.

Frequency Counter

Frequency counters are used to adjust a device's oscillators to a precise, accurate frequency, or to verify a frequency. Radio and TV receivers, video recorders and even some all-digital devices can require carefully set oscillators for proper operation. Frequency measurement also may aid in troubleshooting optical disc players.

A counter works by totaling up how many cycles of an incoming waveform go by in a period of time specified by the instrument's *gate period*. The gate opens, the waves go by, it counts 'em and puts the count on the display. That's it.

Ah, if only real life were that simple! Sometimes this works and gives you a correct count; sometimes it doesn't. For one thing, how does the counter know when a cycle has occurred? Unlike the trigger on a scope, the counter's trigger is very simple: it looks for *zero crossings*, or places where the signal goes from positive to negative, and counts every two of them as one cycle. For simple waveforms with little or no noise, that works great.

A lot of signals have noise or distortion on them, however, that can confuse the zero-crossing detector, resulting in too few or too many counts. If you connect a counter to a test point using a scope probe and then switch between 10X and 1X on the probe, the count will probably change a little bit, with a lower count in the 10X position. Which one is the truth? It's hard to say for sure. Usually, the count is more accurate when the input voltage is lower because noise on the signal is less likely to trip the zero-crossing detector. If it gets too low, though, the detector may miss some cycles altogether, resulting in an incorrect, low count. If you get counts that seem very low for what you were expecting—say, a 10-MHz oscillator reads 7.2 MHz—the input voltage is likely too low and the detector is missing some cycles. If the count seems too high, the signal may be noisy and also could be too strong, adding false cycles to the count.

Complex, irregular signals like analog audio, video and digital pulse trains cannot be counted in any useful way with a frequency counter. What comes in during each gate period will vary, so the display won't settle down. Also, correct tripping of the zero-crossing detector is impossible. Use this instrument only for simple, repeating waveforms such as those produced by oscillators.

Back in Chapter 2, we looked at precision and accuracy, and how they affected each other. Nowhere does this issue come up more often than with frequency counters. Most counters have lots of digits, implying high precision. Accuracy is another matter.

The count you get depends on how long the gate stays open. That is controlled by a *frequency reference*, which is an internal oscillator controlled by a quartz crystal. In a very real sense, the counter is comparing the incoming signal's frequency to that of the instrument's crystal, so variation in the crystal oscillator's frequency will skew the count. Crystals are used in many oscillator applications requiring low drift, but they do wander a bit with temperature and age. Most counters have internal trimmer capacitors to fine-tune the crystal's frequency, but setting them requires either another, trusted counter for comparison; an oscillator whose frequency is trusted; or some cleverness with a shortwave radio that can receive WWV, the National Bureau of Standards atomic clock's time signal originating from Fort Collins, Colorado. That station broadcasts its carrier at a highly precise, accurate frequency, and it is possible to compare it audibly to your counter's oscillator, using the radio as a detector. When they *zero-beat*, or mix without generating a difference tone, your counter is spot on frequency. Performed very carefully with a counter that's been fully warmed up, zero-beating against WWV can get you within 1 Hz, which is darned good.

On many counters, the gate time can be selected with a switch. Longer gate times give you more digits to the right of the decimal point, but their accuracy is only as good as the counter's reference oscillator. Don't take them terribly seriously unless you are certain the reference is correct enough to justify them. If your right-most digit specifies 1 Hz but the reference is 20 Hz off, what does that digit mean? For audio frequencies, you'll need a fairly long gate time to get enough cycles to count. For radio frequencies, a faster gate time is more appropriate.

Connecting a counter to the circuit under test can be tricky in some cases. Loading of the circuit's source of signal generation is a real problem, pulling it off-frequency and affecting the count significantly. Especially when you touch your probe directly to one lead of a crystal in a crystal oscillator, the capacitance of the probe and counter can shift the oscillator's frequency by a surprising amount. To probe such a beast, look for a *buffer* between the oscillator and everything else, and take your measurement from its output. Since the circuit itself also may load the oscillator, it's highly likely that a buffer is there someplace. These days, most crystal oscillators are formed with IC chips, rather than transistors, with the buffer being on the chip and the frequency output coming from a pin *not* connected to the crystal. Use your scope to find it and then connect the counter to that pin.

Analog Meter

Using an analog VOM or FET-VOM requires interpreting the position of a meter needle, rather than just reading some numbers off a display. Why would you bother with this? That meter needle can tell you some things a numerical display can't. Specifically, how it moves may give you insight into a component's or circuit's condition.

Taking most kinds of measurements with a VOM is pretty much like taking them with a DMM. VOMs are not autoranging, so you have to match the selector knob's scale to the markings on the meter movement. Also, you need to zero the ohms scale with the front-panel trimmer knob every time you change resistance ranges. Select the desired range, touch the leads together and turn the trimmer until the meter reads 0 ohms.

Unlike FET-VOMs and VTVMs, VOMs have no amplification, so they load the circuit under test much more when reading voltage than do other instruments. This is of no consequence when measuring the output of a power supply, but it can significantly interfere with some small-signal circuits.

VOMs can pull a few tricks not possible with their digital replacements. Slowly changing voltages will sway the meter needle in a visually obvious way, instead of just flashing some numbers. A little noise won't affect the reading, either, thanks to the needle's inertia. A lot of hum on a DC signal can vibrate the needle in a very identifiable manner. Some old techs could read a meter needle almost as if it were a scope!

If you don't have a capacitance meter, you can gauge the condition of electrolytic capacitors with the meter's ohms scale. This works pretty well for caps of about 10 µf or more; it doesn't work at all for anything under 1 µf or so. Set the meter to its highest range and touch the test leads together. When the needle swings over, use the trimmer on the front panel to adjust it to read 0 ohms. (If the needle won't go that far, the meter needs a new battery!)

Connect the leads to the *discharged* cap, + to + and – to –, and watch what happens. The meter should swing way over toward 0 ohms and then gradually fall back toward infinity. The greater the capacitance, the harder the needle will swing, and the longer it'll take before it finally comes to rest. If the needle doesn't fall all the way back, the cap is leaky. If it doesn't swing toward zero, it's open or of low capacitance.

The meter applies voltage from its battery to the capacitor, so be sure the cap's voltage rating is higher than the battery voltage or you could damage the cap. Some meters use 9-volt batteries for their higher resistance ranges. If yours does, it's wise not to try this test on caps rated lower than that. If your meter uses only an AA cell, there's nothing to worry about.

The ohms scale can also be used to check diodes and some transistors, just as with a DMM.

Contact Cleaner Spray

Cleaner spray is handy stuff, especially for use with older, analog equipment. Volume controls, switches and sockets can all benefit from having dirt and oxidation cleaned away. When controls exhibit that characteristic scratchy sound, it's time to get out the spray.

For spray to be effective, you have to be able to get it onto the active surface of the control or switch. Sometimes that's not easy! If you look at the back of a *potentiometer*, or variable resistor, you may find a notch or slot giving you access to the inside, where the spray needs to be. Always use the plastic tube included with the spray can! Trying to spray the stuff in with the bare nozzle will result in a gooey mess all over the inside of your gear. Once you get some spray into the control, rotate it back and forth a dozen times. That'll usually cure the scratchies.

Switches can be a bit tougher. Some simply have no access holes anywhere. If you can't find one, you'll have to spray from the front, into the switch's hole. Never do this where the spray may come into contact with plastic; most sprays will permanently mar plastic surfaces.

Trimmer capacitors should not be sprayed. They have plastic parts easily damaged by the spray and may lose function after contact with it.

Do your best not to splatter spray onto other components. Wipe it off if you do. Also, it could shatter a hot lamp, so don't use it near projector bulbs. And, of course, avoid breathing it in or getting it in your eyes. Spray has a nasty habit of reflecting back out of the part you're blasting, right into your face. Keep your kisser at least 12 inches away. Farther is better.

To avoid making a mess or getting soaked, try controlling the spray by pressing gently on the nozzle until only a gentle mist emerges. Some cans have adjustable spray rates, but many don't. Some brands are more controllable than others, too. Experiment with this before you attack expensive gear.

Component Cooler Spray

Cooler spray can be incredibly useful for finding thermal intermittents. If a gadget works until it warms up, or it works only *after* it warms up, normal troubleshooting methods can be hard to implement, especially in the second case. How are you going to scope for trouble in something that's working?

Before hitting parts willy-nilly with cooler, decide what might be causing the trouble. The most likely candidates for cooler are power supply components like transistors and voltage regulators, output transistors and other parts with significant temperature rises during normal operation.

As with cleaner, use the spray tube, and try controlling the spray rate. Also, the same caveats regarding breathing it in and getting hit in the face hold here. This stuff is seriously cold and can damage skin and eyes. A small amount hitting your hands won't do you any harm, but I shudder to think of a single drop's splattering on your cornea.

If spraying a suspected component reverses the operational state of your device (it starts or stops working), you've most likely found the trouble.

After spraying, moisture will condense on the cold component. Be sure to kill the power and wipe it off.

Chapter 7

What Little Gizmos Are Made of: Components

Although there are hundreds of types of electronic components and thousands of subtypes (transistors with different characteristics, for example), a small set of parts constitutes the core of most electronic products. Let's look at the most common components and how to recognize and test them. In this chapter, we'll cover out-of-circuit tests, the kind you perform after removing the part from the board. We'll get to in-circuit testing in Chapter 11, when we explore signal tracing and diagnosis.

Capacitors

Capacitors consist of two plates separated by an insulator. A charge builds up on the plates when voltage is applied, which can then be discharged back into the circuit. Capacitors come in many types, including ceramic, electrolytic, tantalum, polystyrene (plastic) and trimmer (variable). See Figure 7-1.

Symbols ⊣⊢ ⊣⊢+ ⊣⊬

Markings

Most capacitors are marked in straightforward manner, with numbers followed by µf (microfarads, or millionths of a farad) or pf (picofarads, or trillionths of a farad). Leading zeros to the left of the decimal point are not shown, so we won't use them here either. Some European gear has caps marked in nf (nanofarads, or billionths of a farad). Thus, a cap marked *1 nf* = .001 µf. Some capacitors are marked with three numbers, with no indication of µf, nf or pf. With these, the last number is a multiplier, indicating how many zeros you need to tack on, starting from picofarads. For instance, a cap marked *101* is 100 pf, because there is one zero after the two numbers indicating

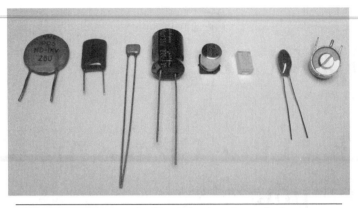

FIGURE 7-1   Ceramic disc, plastic, electrolytic,
surface-mount electrolytic, tantalum, and trimmer
capacitors

the value. A marking of *102*, then, is 1000 pf, or .001 µf. And *103* is 10,000 pf or .01 µf,
and so on. Here's a handy list:

- XX0 = less than 100 pf. The two Xs are the value in pf. Sometimes there's no zero
  on the end. A tiny cap marked *27* is 27 pf. One marked *270* is also 27 pf.
- XX1 = value X 10 pf
- XX2 = .00XX µf
- XX3 = .0XX µf
- XX4 = .XX µf
- XX5 = X.X µf
- XX6 = XX µf

Any value greater than these will be marked directly in µf, as in "2000 µf."

On ceramic disc capacitors, you may also see a marking like *N750*. This specifies
the temperature coefficient, or how much the capacitance drifts with temperature and
in which direction. Keep an eye out for *NP0*, which means no drift in either direction.
NP0 caps are used in time constants and tuned circuits so they won't change frequency
as the unit warms up. Should you ever need to replace an NP0 cap, be sure to use the
same type.

Polarized capacitors are marked for their polarity. With can-style electrolytics, the
marking is a long arrow or black line, and it indicates the *negative* lead. Some very old
electrolytics may show a + sign instead, indicating the positive lead, but they haven't
been made that way for many years. Look for them in antique radios and such.

Tantalum electrolytics, which look like little dipped candy drops with wires, denote
the positive lead with a + sign, or sometimes a red or silver dot.

Surface-mount electrolytics of the can type are marked with a black line or semicircle, usually on the top, indicating the negative lead. Flat plastic caps have one end painted silver or white, and this denotes the *positive* terminal; don't confuse it with the negative-indicating line on can-style caps. Some small surface-mount caps of well under 1 µf have no markings at all, making it impossible to determine their values without a capacitance meter. They are usually tan and are neither polarized nor electrolytic.

If you find a can-type capacitor with no polarity marking, look for *NP*, which indicates a non-polarized electrolytic cap. These are uncommon, but you sometimes run across them in audio gear. You must replace NP caps with the same type.

Uses

Different styles of capacitors cover various ranges and are used for different purposes. Here are the common ones:

- **Ceramic**   These cover the very small values, from a few pf up to around .1 µf, and are used in resonant radio circuits and bypass applications.
- **Plastic**   These start at around .001 µf (1000 pf) and may go as high as .47 µf or so. They are used for bypass and coupling, and are sometimes found in time constants because of their excellent stability over a wide range of temperatures.
- **Electrolytic**   These start at around .47 µf and cover the highest ranges, on up to tens of thousands of µf. They are used for coupling and filtering.
- **Tantalum**   These range from .1 µf to around 47 µf and are a special type of electrolytic capacitor with lower impedance at high frequencies. They are used in filtering and bypass applications when high frequencies are present.
- **Trimmer**   These range from the low pfs to around 200 pf and are used as frequency adjustments for tuned (resonant) circuits and oscillators.

What Kills Them

Different styles of capacitors fail for different reasons, depending on how they're used and to what conditions they're subjected. Generally, application of a voltage above the cap's ratings can punch holes in the dielectric (insulating) layer, heat can dry or crack them, and some wear out with age.

Ceramic and Plastic   These very rarely fail. In all my years of tech work, I've found two bad ceramics and one bad plastic cap! It just doesn't happen. If these caps appear unharmed, they are almost certainly okay.

Electrolytic   These are the most failure-prone components of all. Part of their charge-storing layer is liquid, and it can dry out, short out, swell and even burst the capacitor's seals and leak out. Heat, voltage and age all contribute to their demise. The constant charging and discharging as they smooth ripple currents in normal operation gradually wears the caps out. Today's predominance of switching power supplies, with their fast pulse action, has accelerated capacitor failure. And application of even a little reverse-polarity current will wear out 'lytics in a hurry. A leaky rectifier will permit some AC to hit the filter caps connected to it, resulting in ruined caps. If you replace them without changing the rectifier, the new parts will fail very quickly.

Failure modes include shorts, opens, loss of capacitance from age or drying out, electrical leakage (essentially a partial short), and increased internal resistance, or equivalent series resistance (ESR). If you see a bulge in the top of the cap, or anywhere on it, for that matter, it is bad and must be replaced. Don't even bother to check it; just put in a new one. Look at the bowed top of the capacitor in Figure 7-2. Keep in mind that a cap can also exhibit high ESR or decreased capacitance with no physical signs.

Many surface-mount electrolytics made in the 1990s leaked, thanks to a defective electrolyte formula. If the solder pads look yellow or you see any goo around the cap, the part has leaked and must be replaced. Sometimes the yellow pads are the only clue.

Tantalum   These use a solid electrolyte in the dielectric layer that is quite thin. Consequently, they are prone to shorts from even momentary voltage spikes exceeding their voltage ratings and punching holes in the layer. And they are even less tolerant of reverse current than are standard electrolytics.

Trimmer   Trimcaps use a plastic or ceramic insulating layer that is very reliable. They rely on a mechanical connection, though, between the rotating element and the base, making them prone to failure from corrosion over a period of many years. Sometimes rotating the adjustment through its range a few times can clear it up, but doing so

FIGURE 7-2   Bulging electrolytic capacitor with greatly reduced capacitance

causes loss of the initial setting, requiring readjustment afterward. Never spray the plastic variety with cleaner spray, as it may damage the dielectric layer and destroy the component.

Out-of-Circuit Testing

To test a capacitor after removal from the circuit, use your cap meter. If you don't have one, you can use your DMM's ohms scale to check for shorts. Small-value caps will appear open whether they are or not; they charge up too quickly for you to see the voltage rising.

For electrolytics, do a quick test with an analog VOM, watching for the initial needle swing and slow release back toward infinity. Make sure the cap is discharged before you try to test it. With a polarized cap, connect the test leads + to + and – to –. Most polarized caps will survive a reversed test, but tantalums may not; even momentary reversed voltage can short them out.

Crystals and Resonators

Quartz crystals and ceramic resonators are made from slices of quartz or slabs of ceramic material, with electrodes plated on the sides. They exploit the *piezoelectric effect*, in which some materials move when subjected to a voltage and also generate a voltage when mechanically flexed. Crystals are always found encased in metal, while ceramic resonators are usually in yellow or orange plastic. Crystals have two leads, and ceramics may have two or three. See Figure 7-3.

FIGURE 7-3   Quartz crystals and ceramic resonators

Symbols Crystals and resonators use the same symbol unless the resonator features internal capacitors and has three leads.

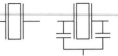

Markings

Crystals and resonators are marked with their frequencies. For crystals, assume the number means MHz. On resonators, assume hundreds of kHz, though some may be in MHz as well. Crystals can sport other numbers indicating the type of cut used, which is quite a complicated topic. It's not a concern in most service work, though. Either the crystal oscillates or it doesn't.

Uses

Quartz crystals are used as frequency-determining elements in oscillators, and sometimes as tuned filters in radio applications. Ceramic resonators are used the same way, but in applications requiring less stability and accuracy, and usually at lower frequencies. You are more likely to find a quartz crystal in the clock oscillator running a digital device like a laptop, DVD player or MP3 player, with a ceramic resonator lurking in a remote control or some radio circuit stages.

What Kills Them

Crystals and resonators are mechanical. They actually move on a microscopic level, vibrating at their resonant frequency. They are also made of crystalline material, so they're somewhat brittle. Heat and vibration can crack them, as can a drop to the floor. Quartz crystals, especially, can develop tiny internal fractures and just quit on their own, with no apparent cause.

Some flaws don't stop them outright; they become finicky and unpredictable. Touching their terminals may cause them to stop or start oscillating. Also, crystals drift in frequency as they age, sometimes drifting past the point at which the circuit will operate properly.

Out-of-Circuit Testing

Without a crystal checker, which is simply an oscillator with an indicator light, there's no way to tell whether a crystal works without scoping its signal in an operating circuit. Even a crystal checker may lie to you, indicating a good crystal that still won't start in the circuit for which it's intended.

Crystal Clock Oscillators

Crystal clock oscillators are complete clocking circuits in a small metal box with four or six pins. Figure 7-4 shows a really small one. (You can see a much larger unit in Figure 10-8, in Chapter 10.) On four-pin parts, two pins are for power and ground,

FIGURE 7-4   12-MHz crystal clock oscillator

a third for output, and the fourth, called *output enable*, to activate or inhibit the output. Six-pin versions sport two complementary outputs (one is high while the other is low), output enable, power, ground and one unconnected pin.

Symbols

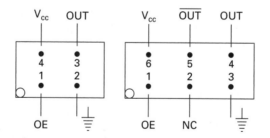

Markings

The frequency will be marked on the case. You may also see a manufacturer's part number.

Uses

More and more, crystal clock oscillators are replacing separate crystals, especially in products using multiple frequencies for various tasks. Because the oscillator is in the can, no extra circuitry is required, so cost and required space are reduced.

What Kills Them

These oscillators contain quartz crystals, so they're subject to the same mechanical issues inherent in crystals. Because the cans also include a complete circuit, they're vulnerable to heat and overvoltage failures as well. For the most part, though, crystal can oscillators are very reliable.

Out-of-Circuit Testing

Applying power and ground to the appropriate pins should produce an output waveform at the frequency shown on the can. The amplitude should be fairly close to the applied DC voltage. Unless no power is getting to the oscillator, it's just as easy to test one in-circuit as out. If there's power, the waveform should be there. The output enable pin must not be low, or no output will appear. It's fine for it to be left unconnected, and in many circuits it is, since the oscillator needs to run all the time anyway.

Diodes

Diodes are one-way valves. Current can flow from their cathodes (–) to their anodes (+) but not the other way. They are made from silicon slabs "doped" with materials that cause the one-way current flow, with two dissimilar slabs touching at a junction point. Large diodes used in power supplies are called *rectifiers*, but they do the same thing. Two of them in one package, sharing one common terminal (for a total of three leads), are called a *double diode,* or *double rectifier.* Four of them arranged in a diamond-like configuration are called a *bridge rectifier,* whether they are separate parts or integrated into one package. See Figure 7-5.

FIGURE 7-5 Diode, rectifier, and bridge rectifiers

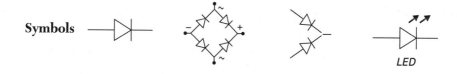

Symbols

LED

Markings

Single diodes and rectifiers have a band at one end indicating the cathode. They are marked by part number, which does not indicate their operating parameters like maximum forward and reverse operating voltages and maximum current. To get those, you must look up the part number in a book or online.

Bridge rectifiers may be marked for peak forward voltage and current, but they often have part number markings or none at all. Sometimes you'll see " ~ " at the AC inputs and " + " and "–" at the DC outputs.

You may see a marking like *200 PIV*, for peak inverse voltage. This is the maximum voltage the diode can withstand in the reverse (nonconducting) direction before breaking down and allowing the voltage to pass. Exceeding the PIV rating usually destroys the diode.

Uses

Diodes and rectifiers are common parts in pretty much every electronic product. Small-signal diodes rectify signals for detection of the information they carry, as in a radio, and direct control voltages that turn various parts of the circuitry on and off. They may also be used as *biasing* elements, providing a specific current to the inputs of transistors and other amplifying elements, keeping them slightly turned on so that they can conduct over the required portion of the input signal's waveform. Light-emitting diodes, or LEDs, are used as indicators on control panels, and bright white LEDs backlight newer LCD screens in many products, from laptops, netbooks, and tablet computers to large TVs.

Rectifiers convert incoming AC power to DC. They convert the pulses in a switching power supply's transformer back to DC as well. Bridge rectifiers are commonly used to change AC line current to DC by directing opposite sides of the AC waveform to the appropriate + and – output terminals.

What Kills Them

Diodes and rectifiers can fail from voltage exceeding their limits, but the most common cause is too much current and the heat it produces. Sometimes they just fail with age, too. Failure modes include opens, shorts and leakage.

Out-of-Circuit Testing

Most DMMs include a diode test function that shows the voltage drop across the part, indicating whether it is passing current. Be sure to test the diode in both directions. It should show a drop of around 0.6 volts in one direction and appear open in the other. An open reading will show the applied test voltage, the same as when the meter's leads are unconnected to anything. Even a small voltage drop in the reverse direction suggests a leaky diode that should be replaced.

With an analog VOM, you can use the resistance ranges to get a good idea of a diode or rectifier's condition, but small leakage currents are hard to detect. With either type of meter, you're most likely to see a total failure, either open or shorted, if the part is bad. Leaky diodes are rare, but not so rare that you shouldn't keep the possibility in mind.

Fuses

Fuses protect circuitry and prevent fire hazards by stopping the current when it exceeds the fuses' ratings. Though simple in concept, fuses have a surprising number of parameters, including current required to blow, maximum safe voltage, maximum safe current to block and speed of operation.

The primary parameter is the current required to melt the fuse's internal wire and blow it. If you're not sure of anything else, be sure to get that right when replacing a fuse.

The speed at which the fuse acts is also important in some devices. Time-delay or *slow-blow* fuses are used for applications like motors, which may require momentary high start-up current. Ultra-fast-acting fuses are used with especially sensitive circuitry that must be protected from even transient overcurrent conditions. Most consumer electronic gear uses standard fuses, which are considered fast-acting but not ultra-fast.

Fuses come in many shapes and sizes, from the ubiquitous glass cylinders with metal end caps to tiny, rectangular, surface-mount parts hardly recognizable as what they are. You'll find fuses in holders and also soldered directly to circuit boards. Be on the lookout for glass fuses with internal construction including a spring and a little coil; those are the slow-blow variety. See Figure 7-6.

Symbol   —o_o—

Markings

At least four marking systems are used on fuses. The primary marking is the melting current, shown by a number followed by an *A*. You may also see *3AG* or *AGC*, both of which indicate standard-speed glass fuses. You must look up other markings to get the speed rating. Many online catalogs offer this information.

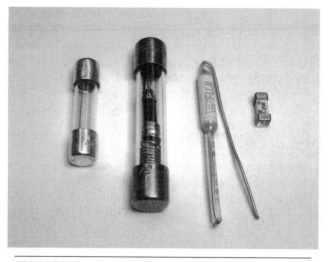

FIGURE 7-6   Fuses

Uses

Anything that plugs into the wall will have a fuse on the hot side of the AC line, usually right as it enters the device, before it even gets to the power switch. Battery-operated gadgets often have fuses too, between the battery's positive terminal and the rest of the unit. Some products have multiple internal fuses protecting various parts of the circuitry. I've seen as many as eight of them in one camcorder!

Some audio amplifiers and receivers use fuses in line with the speakers to protect the amp, should a speaker's voice coil overheat and short out. Such a fuse can also protect the speaker if the amp develops a shorted output transistor and sends the power supply's entire current capacity straight to the speaker.

What Kills Them

Most fuse failures are caused by doing their job. A short in the circuitry pulls too much current through the fuse, so it blows. Now and then, you may run across a fuse that has fatigued with age and use, finally failing. If it's a glass fuse, take a look at the inside. When the two wires are almost touching and there's no discoloration on the glass, the fuse blew gently, and there's a possibility that the circuitry isn't shorted. That happens sometimes with speaker fuses when the amplifier is played at high volume for extended periods. The fuse's wire gets just warm enough to fracture, but there's no malfunction in the circuitry. If you see a wide gap between the wires and spattering on the inside of the glass, the fuse blew hard, indicating a lot of current and certain circuit failure.

Very rarely, fuses can develop resistance, continuing to pass current but interfering with the full flow. I've seen it a couple of times, and my first such case drove me bonkers trying to figure out why the power supply voltage was low and erratic. Figuring a fuse was either good or bad, I never considered it as a possible culprit until I'd wasted hours looking at everything else. Incredulous, I pulled it and discovered that it had become a 10-ohm resistor! I haven't trusted the little buggers since.

Out-of-Circuit Testing

Check fuses with your DMM's ohms scale. It should read 0 ohms. If you see an ohm or 2, try touching the meter's leads together; you may get the same reading, thanks to the resistance of the leads themselves. If not, and the resistance is definitely in the fuse, replace it. A blown fuse will read completely open, of course.

Inductors and Transformers

Inductors, or coils, generate a magnetic field when current passes through them. When the current through the inductor stops or changes direction, the field collapses and creates a current in the wire, opposing the changes. The effect is to store some of the energy and put it back into the circuit.

Coils may be wound on nonferrous cores having no effect on the magnetic field, or they may be wound on iron cores that play a significant role in the inductor's behavior. Two inductors wound on the same iron core can be used to convert one voltage and current to another by creating a magnetic field in one coil and using it to generate a current in the other coil, which may have a different number of turns of wire, thus creating a different voltage. This arrangement is called a *transformer*. See Figure 7-7.

FIGURE 7-7   Inductors and transformers

Symbols

Markings

Many coils are not marked. Some small ones are encapsulated in plastic or ceramic and marked in μh (microhenries, or millionths of a henry) or mh (millihenries, or thousands of a henry).

Uses

Inductors and transformers are used to convert voltages; to couple signals from one stage to another, especially in radio equipment; and to filter, or smooth, voltage variations. They are also essential parts of many resonant circuits, working together with capacitors to establish *time constants*.

Their inherent opposition to rapid voltage change makes coils useful for isolating radio-frequency signals within circuit stages while allowing DC to pass. When used in this manner, coils are called *chokes* because they choke off the signal. Look for them where DC power feeds from the power supply into RF stages.

What Kills Them

In low-power circuits, coils are highly reliable. Failure is pretty much always due to abuse by other components. It's very rare for small-signal inductors to fail, since little current passes through them. In circuits where significant supply current is available, a shorted semiconductor can pull enough through the inductor feeding it to blow the coil. If you find an open inductor, assume something shorted and killed it.

The windings in power transformers sometimes open when too much current overheats them and melts the wire. Often, the primary (AC line) side will burn out even though the excessive current draw is on the other side of the transformer, somewhere in the circuitry. They may also short to the iron core. The insulation of the windings in high-voltage transformers can break down and arc over to other windings, create a short between windings, or arcing and shorts to the core.

Out-of-Circuit Testing

Use your DMM to test for continuity from one end of a coil or section of a transformer to its other end. You can also use the ohms scale to check for shorts from windings to the core and between unconnected windings.

It's very hard to tell whether windings are shorted to each other in the same coil. Using the ohms scale, the difference can be so slight that it's undetectable. If you have an inductance meter and know the correct inductance, that will give you some idea. With unmarked coils and just about all transformers, though, you won't know what the inductance should be without access to an identical part for comparison. In some circuits, such as LCD backlighting inverters, there can be two identical transformers, so you may be in luck.

Integrated Circuits

ICs, or chips, come in thousands of flavors. In today's advanced products, ICs do most of the work, with transistors and other components supporting their operation. Digital chips are at the hearts of computers, DVD players, digital cameras, MP3 players, you name it. In many devices, they work side-by-side with analog ICs handling radio, audio and video signals. See Figure 7-8.

ICs integrate anywhere from dozens to millions of transistors on a small square of silicon, with microscopic structures printed using photographic techniques. A failure of even a single transistor will render the chip defective. It's pretty amazing that they ever work at all! Obviously, there's no way to test the individual structures; all you can do is verify whether the chip is properly performing its intended function.

There are some off-the-shelf chips used in many products, but custom ICs specific to a model or product category are quickly coming to dominate our gadgets' innards. Each chip can include more product-specific functions, so it takes fewer of them to make a device. The fewer parts and interconnections, the more reliable an item is likely to be. And that gadget can be smaller and cheaper to build.

Large-scale-integrated chips, called LSIs, can have up to a few hundred pins spaced so closely that you can't even put a probe on one without shorting it to its neighbor. Without exceedingly expensive, specialized equipment, it's very difficult to unsolder or

FIGURE 7-8 Small-scale and large-scale ICs in a DVD player

resolder these parts. There is a trick for touching up suspected bad joints on some of them, but it takes practice and doesn't always work. We'll look at it in Chapter 12.

Luckily, ICs are very reliable. Most of them handle small signals, so they don't dissipate a lot of power and get hot enough to fail. There are some exceptions to that, however—notably CPUs and video graphics chips in computers. Some graphics cards have fans over the chips, as do CPUs. You know the thing gets mighty warm when it needs its own fan! These parts get so hot because they have millions of transistors switching from millions to a few billion times per second. Those microscopic heat generators add up to a serious temperature rise.

Other hot-running chips include audio power output modules in stereo receivers, some types of voltage regulators, motor controllers, convergence chips in big-screen CRT TVs and anything else that pumps real power. These parts are usually mounted on heatsinks, and they fail as often as power transistors.

Symbols   Chips are denoted on schematics by the number of pins and their general shape. Some simple chips, like logic gates and op amps, may include a schematic of their general internal functions, but not the actual layout of transistors on the chip.

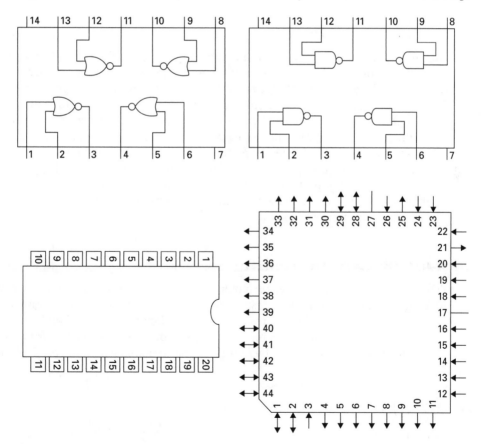

Markings

The parts are marked with part numbers that can mean many things. Small-scale, industry-standard chips are produced in *families* sharing some part number commonality. For instance, older CMOS logic gate chips are called "4000 series" and have part numbers like 4011B and 4518. They all begin with *4*.

Large-scale integrated circuits often have proprietary part numbers, and they certainly will if they're custom parts made for the specific type of product.

The pins on an IC are numbered going counterclockwise around the chip in a ring. Pin 1 will have a dimple or spot next to it on the chip's plastic casing, and it will be in a corner.

Uses

ICs are used for just about everything: audio amps, data processors, logic gates, oscillators, signal processing, and any other function you can think of.

What Kills Them

ICs are highly reliable, but heat is a major danger, especially when it's internally generated. Voltage spikes can destroy some chip families, but more modern varieties are fairly voltage-tolerant. Still, a static charge may present a voltage too high for any chip to withstand. A short in another area of the circuit that pulls too much current from a chip's output can blow it. If the power supply feeding the chip fails but signals are still fed to it, the chip can *latch*, causing a permanent internal short. Finally, some ICs, especially LSIs, have internal features so small that atomic forces may gradually eat through the microscopic wires and connections, creating holes that wreck the chip. Essentially, the part dies of old age. This problem has been researched by chip makers for many years, and today's ICs hold up rather well. Now and then, a chip still dies without apparent provocation.

Out-of-Circuit Testing

There's no way to put a meter on a chip to see if it's good. In fact, doing so may damage the chip if the meter's test voltage happens to be high enough and touches the wrong pins. Simple logic gates can be plugged into a test board and hooked up to test their functions, but the exercise is more trouble than it's worth. For the most part, chips must be tested in-circuit by observing their actions with your oscilloscope.

Op Amps

The op amp, or operational amplifier, is a common type of chip. It's an analog, general-purpose amplifier configurable to do many different jobs, such as buffering,

FIGURE 7-9 Quad op amp

oscillating and signal filtering, by connecting resistors and capacitors to its inputs and outputs. The number of op amp circuit variations is staggering!

One chip may contain several op amps. Dual and quad op amps are used in many products. See Figure 7-9.

Symbol

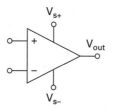

Markings

The chips are marked with part numbers, and there are some, like LM358, you'll see often.

Uses

You'll find op amps in audio circuits, radio circuits, motor controllers and power supplies. Any place requiring an amplifier is a prime candidate for an op amp. Most op amps are for small-signal applications, but some power op amps are capable of driving significant loads. Expect those to be heatsinked.

What Kills Them

Heat and overcurrent are the primary culprits.

Out-of-Circuit Testing

As with logic gates, op amps have to be in some kind of circuit for you to test them.

Resistors

Resistors offer opposition to current, dissipating the opposed power as heat. Limiting current is a vital function in any circuit, so every electronic product has resistors. They come in many shapes and sizes, but those with leads are easily identifiable by their color bands indicating the resistance value in ohms. Some large resistors are marked numerically, as are some of the tiny surface-mount parts. See Figure 7-10.

The basic type of resistor, found in virtually everything, is the carbon composition resistor. Made from a carbon compound, these resistors run the full range of values, from less than 1 Ω (ohm) to 10 MΩ (megohms, or millions of ohms). Most you'll see will be over 10 Ω and under 1 MΩ. The tolerance for standard carbon comp resistors is ± 5 percent. That is, the measured value should be no more than 5 percent high or low of the stated resistance.

Some applications require the use of wire-wound resistors. These look a lot like carbon comp parts, but you can see the coil of wire under the paint on the body. Wire-wound resistors can dissipate more heat than can carbon comps. They also can be manufactured to very tight tolerances, but they have some inductance, since they

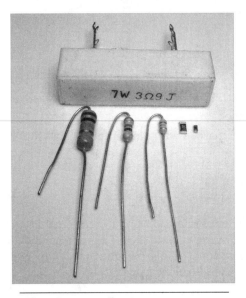

FIGURE 7-10 Resistors

are coils, and are ill suited to high-frequency circuits where the inductance might matter. The two types should not be interchanged.

In addition to the resistance value, the power dissipating capability of a resistor, measured in watts, is an important factor. Since power is dissipated as heat, knowing how much heat the part can take before disintegrating is vital. Standard carbon comps are rated at 1/4 watt, with some slightly larger ones able to dissipate 1/2 watt. Very tiny ones with leads are rated at 1/8 watt, while surface-mount versions typically vary from 1/8 watt down to 1/32 watt.

Other resistor formulations include carbon film, metal film and metal oxide. Carbon film types introduce a bit less noise into the circuit and are used in areas where that's a significant issue. Metal film and metal oxide parts have tighter tolerances, in the range of 1 or 2 percent of stated value. Some circuits require that for proper operation.

Symbol ─╱╲╱╲╱╲─

Markings

The use of color-coding dates back to the vacuum tube days, when resistors got so hot that printed numbers would evaporate. To read the color code, you must first determine which end of the resistor is the start and which is the far end. Look for a gold or silver band; that's the tolerance marking, indicating the far end, and there's usually a little extra space between that band and the others. The first digit will be at the end farthest from the tolerance band.

Each number is represented by a color. The scheme is as follows:

Black	0	Green	5
Brown	1	Blue	6
Red	2	Violet	7
Orange	3	Gray	8
Yellow	4	White	9

The tolerance bands at the far end are as follows:

Brown	± 1 percent
Red	± 2 percent
Gold	± 5 percent
Silver	± 10 percent

To determine a resistor's value, read the first two bands as numbers. The third band is a multiplier, indicating how many zeros you need to tack on to the numerical value. So, for instance, red-red-brown would be 2, 2, and one zero, or 220 ohms. Yellow-violet-orange would be 4, 7, and three zeros, or 47,000 ohms, a.k.a. 47 KΩ. Be careful not to confuse a black third band with a zero; it means *no* zeros.

This scheme works great until the resistance value gets below 10 Ω. Try it—there's no way to mark such a value. For resistors that low, the third band is set to gold, meaning that there's a decimal point between the first two bands. So, green-blue-gold-gold would be a 5.6 Ω resistor with a ± 5 percent tolerance. You won't see too many resistors under 10 Ω, but you might run across one in an audio output stage or a power supply.

Some resistors have four bands plus a tolerance band. These are higher-precision parts with tighter tolerances and must be read slightly differently. With these, read the first *three* bands as numbers and the fourth band as the multiplier. So, the 47 KΩ resistor would read yellow-violet-black-red.

How can you tell what type of resistor you have? Sometimes you can't. If the resistor has a gold or silver tolerance band, assume it's a standard carbon composition part, unless it's in a very low-noise circuit like an audio preamp, in which case carbon film might be a more appropriate replacement. If it has a red or brown tolerance band, indicating higher precision, it might be a metal film or metal oxide component.

Many surface-mount resistors are marked numerically, using the same idea. The last number is a multiplier. If you see an *R* between the numbers, that's a decimal point. You'll see that only on resistors of rather low value. For instance, 4R7 means 4.7 ohms. If you see a number with a zero at the end, don't confuse that to mean a number; it indicates *no* zeros. Thus, 220 means 22 Ω, not 220 Ω, and 220 Ω would be marked 221. Also, a letter *after* the Ω on any type of numerically marked resistor is not part of the numerical value, even if it's a *K*. So, 47 ΩK is still 47 Ω, not 47 KΩ. Used this way, the *K* denotes 10 percent tolerance. Pretty crazy, huh?

Some tiny surface-mount resistors are too small for numerical value markings. Instead, they sport a two-digit number that must be cross-referenced from a list. The number itself has no direct relation to the value. This marking scheme has several permutations, and you can find them on the Internet. It's highly unlikely, though, that you'll ever need to replace one of those tiny resistors, because they carry very little current and rarely fail.

Uses

Resistors are found in just about every circuit. They limit the current that can pass through other parts. For instance, transistors amplify a signal by using it as a control for a larger current provided by the power supply, somewhat like the handle on a spigot controls a large flow of water. A resistor between the supply and the transistor sets how much current the transistor has to control. Without the resistor, the transistor would have to handle all of the supply's current and would self-destruct.

What Kills Them

Resistors rarely fail on their own. Heat caused by passing too much current burns them out, sometimes literally. Carbon composition resistors can go up in flames or become a charred lump when a short in some other part pulls a lot of current through them. It's not uncommon to see one with a burn mark obscuring the color bands.

Out-of-Circuit Testing

Use your DMM's ohms scale to see if the resistance is within the specified tolerance range. Most resistors do better than their tolerances, but expect a small difference between what's on the code and what you measure.

When the resistor is charred beyond your being able to read the color code, it can be a real problem unless you have the unit's schematic diagram. Luckily, resistance values follow a standard pattern, because there would be no point in producing resistors whose values fell within another resistor's tolerance range. So, you may be able to infer a burned-out band's value from others that can still be seen. If you ever need it, you can look up the list of standard values online.

Potentiometers

A potentiometer, or *pot*, is a variable resistor. A small one used for internal circuit adjustments is called a *trimpot*. Pots and trimpots may have two, three or (rarely) four leads, but most have three. The outer two are connected to the substrate on which the resistive element is formed, with one lead at each end of the resistor. The center lead goes to the wiper, a movable metal contact whose point touches the resistive element, selecting a resistance value that rises relative to one outer lead while falling relative to the other as you turn the knob. See Figure 7-11.

A two-wire pot has no connection to one end of the resistive element but is otherwise the same. Some two-wire pots connect the free end of the element to the wiper, which slightly affects the resistance curve as you turn it, but it doesn't matter a whole lot. Two-wire pots are sometimes called *rheostats*.

FIGURE 7-11   Pots and trimpots

Four-wire pots, used mostly on stereo receivers to provide the "loudness" function which increases bass at low volume levels, are like three-wire pots, but with an extra tap partway up the resistive element.

The amount of resistance change you get per degree of rotation is even from end to end on *linear taper* pots. On *log taper* pots, also called *audio taper*, a logarithmic resistance curve is used, so that audio loudness, which is perceived on a log curve, will seem to increase or decrease at a constant rate as the control is varied. Trimpots, whose primary use is for set-and-forget internal adjustments on circuit boards, are always linear taper.

How much power the pot can dissipate depends on its size. It's not marked on the part. In most applications, only small signals are applied to it, so it's not much of an issue.

Some pots have metal shafts, while others use plastic. Plastic provides insulation in applications like power supplies, where you might come in contact with a dangerous voltage. Also, some pots have switches built into them, and those may be rotary, operating at one end of the wiper's travel, or push-pull.

Most trimpots rotate through less than 360 degrees, just like pots. Those used in applications requiring high precision may be multi-turn, with a threaded gear inside providing the reduction ratio. Multi-turn trimpots use a small screw for adjustment, usually off to one side of the body of the component. Be wary of turning the screw, because there's no visual indicator to help you set it back where it was.

In stereo receivers, pots can be ganged together onto one shaft, so that turning it will affect both channels together. Each pot is internally isolated from the other, though you may find one end of all of them tied to ground at the terminals.

Symbols

Markings

Pots are numerically marked with their resistance values using the same scheme as resistors. A *B* in the code indicates a linear pot, while a *C* means a log pot. An *A*, though, can mean either, as the codes have changed over the years. Some pots have *LIN* or *LOG* printed on them. Most don't, though.

Uses

Pots are used to adjust operating parameters for analog signals and power supply voltages. Once, they were the primary method of setting just about everything. In this digital age, volume, treble, bass, brightness, contrast and such are more often selected from a menu or adjusted with up/down buttons.

Trimpots on circuit boards are less common too, but you'll still find some, especially in power supplies, including switchers.

What Kills Them

Most failures are caused by wear or dirt where the wiper makes contact with the resistive element. The symptom is scratchiness in audio, flashes on the screen in video, or inability to set the pot to specific spots without the wiper's losing contact with the element. Contact cleaner spray will usually clear it up, but if the element is too worn, replacement is the only option.

Occasionally, the resistive element cracks, resulting in some weird symptoms because the wiper is still connected to one end but not the other. Plus, which end is connected reverses as you turn the control over the broken spot. Audio may blast through part of the control's range and disappear below the break point.

Out-of-Circuit Testing

Test the outer two leads on a three-lead pot with your DMM's ohms scale. It should read something close to the printed value. If the element is cracked, it'll read open.

To check for integrity of the wiper's contact, connect the meter between one end of the pot and the wiper. Slowly turn the pot through its range, observing the change in resistance. This is one test better done with a good old analog VOM; the meter needle will swing back and forth with the position of the wiper, and it's easy to interpret.

A two-lead pot should read something close to its stated resistance at one end of its control range and 0 ohms at the other.

To verify if a pot is linear taper or log taper, measure the resistance at one-third of the rotation and again at two-thirds. See if those values are about one-third and two-thirds of the total resistance from end to end. If it's a log taper part, they won't even be close.

Relays

Relays are switches controlled by putting current into a coil. When the coil is energized, the resulting magnetic field pulls a metal plate toward it, pressing the attached switch contacts against opposing contacts. In this way, a small current can control a much larger one, just as a transistor does in *switch mode*.

Relays may have any number of contacts for switching multiple, unconnected circuits at the same time. As with switches, each set is called a *pole*. Also, some contacts may be *normally open*, or *NO*, meaning that they are not touching until the relay is energized, and some may be *normally closed*, or *NC*, meaning the opposite. Each direction is called a *throw*. Thus, a double-pole, double-throw, or DPDT, relay would have two switches, each with three contacts: the NO side, the movable contact, and the NC side. See Figure 7-12.

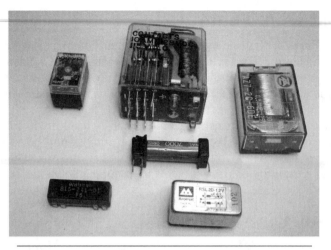

FIGURE 7-12 Relays

Symbols

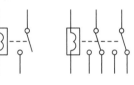

Markings

Relays may be marked schematically for their internal construction—that is, where the coil connections are and what kinds of switches are inside. You may find a voltage rating; that specifies the coil voltage, not the maximum voltage permitted on the switch contacts. A resistance rating is for the coil too. If you find one, you can calculate the coil's pull-in current with Ohm's Law. It also helps you when testing the coil with your DMM.

Some relays include an internal diode across the coil to prevent the reverse voltage it generates when power is removed from feeding back into the circuitry and damaging it. If the relay shows polarity markings (+ and –), it has a diode.

The maximum current the switch contacts can handle usually isn't shown, but it might be. If the markings read *12 VDC, 3 A*, that indicates a 12-volt coil intended to be driven with DC power, with switch contacts capable of switching 3 amps. Some relays are made specifically for AC coil operation, too, with the appropriate markings.

Uses

Relays were once widely used to switch large currents with smaller ones. These days, semiconductors usually do that job, but some applications still employ relays. Power supply delay circuits, which prevent power from reaching the circuitry for a few moments after the supply is turned on, often have relays. Speaker protection circuits in high-power audio amplifiers use them too, because the high-current audio signal is not impeded at all by a relay, but it would be by a semiconductor.

Most relays make an audible click when they switch, giving their presence away.

What Kills Them

Relay troubles usually involve the switch contacts. Corrosion from age and oxidation, and pitting from arcing when large currents are switched, cause resistance or flaky contact. Once in awhile, a relay will become sticky, not wanting to open back up after power is removed from the coil. This condition can be caused by arcing in the contacts making them stick to each other, and by weakening of the spring used to pull the plate away from the coil's iron core.

If the relay has a removable cover, you may be able to pop it off and clean the contacts. Sometimes just pulling a piece of paper soaked in contact cleaner between them will spiff them up and restore proper operation. If that's not enough, very light wiping with fine sandpaper may remove the outer layer of gunk. Silver polish works too, but be sure to get it all off when you're done. Be careful not to bend the contacts, and don't sand off the plating; it's vital for their long-term survival. Whichever method you use, wipe the contacts with cleaner-soaked paper to remove residue before you put the relay's cover back on.

Out-of-Circuit Testing

Use your DMM's continuity or lowest ohms scale. Check for coil continuity. If there's a diode, be sure to check in both directions. The coil shouldn't read 0 ohms; there's enough wire there for dozens to a few hundred ohms. If it reads very near zero on a relay that has a diode, suspect a shorted diode, especially if the symptom suggests that the coil doesn't want to pull in the switch. Also check that there is no continuity between the coil and its metal core. If there is any, you need a new relay.

Check the NC contacts with the meter. They should read 0 ohms or very close to it. Use your bench power supply to energize the coil. If it has a diode, be certain to get the polarity correct, with + to the diode's cathode, not its anode. (Remember, the diode is supposed to be wired backward, so it won't conduct when power is applied.) With no diode, polarity doesn't matter. Once you hear the click, check the NO contacts. They should also read 0 ohms or very close to it. When you disconnect the power supply, the contacts should return to their original state. The NO contacts should open and the NC contacts should close.

Switches

Switches permit or interrupt the passage of current. There are many, many kinds of switches, and they're used in just about everything. Toggle switches, slide switches, rotary switches, leaf switches, pushbuttons, internal switches on jacks...there's practically no end to the varieties. See Figure 7-13. Many newer products do not

FIGURE 7-13 Switches

have "hard" power switches that actually disconnect power to the unit. Instead, a low-current switch signals the microprocessor, which then shuts down power using semiconductors to interrupt the flow.

Switches can have any number of contacts for switching multiple, unconnected circuits at the same time. Rotary and slide switches, especially, may have many sets, while toggle switches rarely have more than two or three. Each set is called a *pole*. Also, some contacts may be *normally open*, or *NO*, meaning that they are not connected with the switch in the "off" position, and some may be *normally closed*, or *NC*, meaning the opposite. Each direction is called a *throw*. Thus, a double-pole, double-throw, or DPDT, switch would have two separate sets of contacts, each with three elements: the NO side, the movable contact, and the NC side.

Symbols

Markings

If marked at all, switches may show their maximum voltage and current ratings.

Uses

Expect to find switches everywhere. From pushbuttons on front panels and remote controls to tiny slide switches on circuit boards, they handle power and information input in essentially all products. Leaf switches, with bendable, springy metal arms, sense the position of mechanisms like tape transports and laser optical heads, informing the microprocessor of the state of moving parts. Switches inside jacks sense when accessories are plugged in, altering system behavior to accommodate them.

What Kills Them

As with relays, age, oxidation and contact pitting from arcing usually do them in. If the switch's construction permits any access, try spraying some contact cleaner inside, and then work the switch a bunch of times.

Out-of-Circuit Testing

Test switches with your DMM's continuity or lowest ohms scale. The contacts should read 0 ohms when closed and infinity when open. There's an exception, though: Some pushbuttons, such as the kind on remotes, laptop keyboards and tiny products like digital cameras, use a carbon-impregnated plastic or rubber contact to make the connection. You can identify them by their soft feel when pressed; they don't click. These switches are intended only for signaling, not for handling significant current, and they may have a few tens of ohms of resistance when in the "on" state. While such a reading would indicate a bad toggle switch, it's fine with these little guys.

Transistors

Along with integrated circuits, transistors are the active elements that do most of the work in circuits, amplifying, processing and generating signals, switching currents and providing the oomph needed to drive speakers, headphones, motors and lamps. See Figure 7-14.

FIGURE 7-14   Transistors

Transistors act like potentiometers, but instead of your hand's turning the shaft, a signal does. The current fed by the power supply through the pot can be much greater than that required to turn the shaft, providing *gain*, or amplification, as the wiggling signal molds a bigger version of itself from the power supply's steady DC.

There are thousands of subtypes of transistors, but most fall into three categories: bipolar, JFET and MOSFET.

Bipolar transistors are the standard types used in products since the 1950s. They come in two polarities, NPN and PNP, and consist of three elements joined by two junctions. The three elements are the *base, emitter* and *collector*, each with its own lead. Current passing between the base and the emitter permits a much larger current to pass between the collector and the emitter, with one of them being fed from the power supply. The greater the voltage difference between the base and emitter, the more current will pass, and a proportionally higher current can pass between collector and emitter. In most configurations, the signal is applied to the base, causing a bigger version of itself to be formed from the flow between the collector and emitter. The ratio of base-to-emitter current to collector-to-emitter current is the transistor's *gain*, and is inherent in the component's design.

In an NPN transistor, the base must be positive with respect to the emitter for collector-to-emitter current to pass. In a PNP transistor, the base must be negative. So, the two types are of opposite polarity and cannot be interchanged. Most transistors used today are NPN, but you will find circuits with some PNP parts.

JFETs, or *junction field effect transistors,* work on a similar principle. They label their three elements differently. Instead of the base, the controlling terminal is called the *gate*. The emitter is called the *source*, and the collector is the *drain*.

Instead of a current passing from gate to source, application of a voltage to the gate goes nowhere but results in an electric field that controls a channel in the transistor, permitting current to pass between drain and source. This gives the JFET a very high input impedance, which is another way of saying that it does not take much signal current to turn it on.

MOSFETs, or *metal-oxide-semiconductor field-effect transistors,* are similar to JFETs, and they use the same terminal names, but their internal construction is a bit different. They have even higher input impedance and some other desirable characteristics that have resulted in their pretty much dominating FET applications. You may see JFETs in older gear, but you're much more likely to see nothing but MOSFETs in newer equipment.

Like bipolars, FETs come in two polarities, P-channel and N-channel, corresponding to PNP and NPN bipolar transistors. They also come in *enhancement mode* and *depletion mode* types, specifying what happens when the gate voltage is zero. An enhancement mode FET will be turned off with no voltage at the gate; like a bipolar transistor, it requires a *bias* voltage to turn it on. A depletion mode FET will be turned on with zero gate voltage. The only way to turn it off is to apply a voltage of opposite polarity to the one that will increase current flow.

Luckily, you don't need to worry too much about these arcane details. If a FET is bad, you'll look up its part number and replace it with a compatible type. Still, knowing the basics of how these parts work is essential for understanding how to troubleshoot circuits using them...which is pretty much all circuits.

As you can see, transistors have many parameters, so it's not surprising that there are thousands of subtypes with different gains, power dissipation capabilities, frequency limits and so on. Some are similar enough that they can be interchanged in many circuits, but most are not. To replace one part number with another, you need a transistor substitution book or an online cross-reference guide.

Symbols

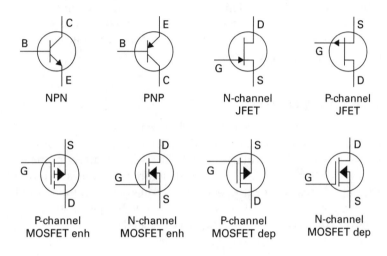

| NPN | PNP | N-channel JFET | P-channel JFET |

| P-channel MOSFET enh | N-channel MOSFET enh | P-channel MOSFET dep | N-channel MOSFET dep |

Markings

Transistors are marked by part number, called a *type number*. There are thousands of these numbers! Some numbers indicate whether a bipolar part is NPN or PNP. If the number starts with *2SA* or *2SB*, it's PNP. If it starts with *2SC* or *2SD*, it's NPN. Sometimes the *2S* will be left off, and there are plenty of type numbers that don't follow this scheme at all. A *3N* indicates a FET, but some of them have numbers starting with *2N*, and there are lots of other kinds of numbers for these too.

Some transistors have *house numbers*, which are proprietary numbering schemes used by different manufacturers to mean different things. There is no way to ascertain what the industry-standard number would be for such a part. Tiny, surface-mount transistors often have no numbers at all.

The arrangement of the leads varies with transistor type. Small Japanese parts with leads are usually laid out ECB, left to right, while American parts are often EBC. Metal-encased power transistors have only two leads and use the metal casing as the collector. Plastic power transistors are usually BCE, with the metal tab, if there is one, being C.

Uses

In discrete (nonintegrated-circuit) stages, transistors are the active elements doing the work, with passive parts like resistors and capacitors supporting their operation. You'll see this kind of construction in radio receivers, audio amplifiers and some sections of many other products. In stages where an IC is at the center of the action, transistors frequently do the interfacing between the IC and other parts of the circuitry, especially areas requiring more current than an IC can supply. MOSFETs are used as switches, permitting the microprocessor to turn power on and off to various parts of the circuitry. Some of their very sensitive varieties are used to amplify and detect radio signals. Many audio power amplifier output stages are made from bipolar transistors. It's hard to find any function that transistors *don't* do. After electrolytic capacitors and bad connections, transistors will be the focus of much of your repair work.

What Kills Them

Transistors are not especially fragile, but they work hard in many circuits and fail more often than do most components. Overheating due to excessive current will burn them out, as will too high a voltage. MOSFETs are particularly prone to shorts from static electricity. Sometimes the internal structure of a transistor develops a tiny flaw, and the thing self-destructs with no apparent cause. In fact, many random product failures occur for precisely this reason. You change the part and the unit works again, and you never find any reason for the dead transistor.

Out-of-Circuit Testing

If you have a transistor tester, use it! Nothing's easier than hooking up the leads and getting the test result. If you don't have one, you can use your DMM's diode test or, lacking that, the ohms scale to check for shorts. If you get near-zero ohms between any two leads, check in the other direction. If it's still near zero, the part is shorted.

Checking for opens is a bit more complicated, because some combinations of terminals *should* look open, depending on to what the control terminal is connected. Connecting the base of a bipolar transistor to its collector should result in its turning on, showing measurable resistance between the emitter and collector in one direction. Connecting the base to the emitter should turn it off.

Similarly, connecting the gate of a FET to its drain should turn it on, and connecting it to the source should turn it off. However, some FETs are symmetrical and will turn on with the gate connected to either of the other terminals, as long as the polarity of the applied voltage is what the gate needs. And the whole situation is complicated by the enhancement/depletion mode issue, because a depletion mode FET will stay on. FETs are not easy to test with an ohmmeter!

Voltage Regulators

Voltage regulators take incoming DC and hold their output voltage to a specific value as the incoming voltage fluctuates or the load varies with circuit operation. Yesteryear's simple products often had no voltage regulators, but practically everything made today does.

Linear regulators act like automatic variable resistors, passing the incoming current through a transistor called the *series pass transistor* and setting the base current to keep the output voltage constant. When the load changes and the voltage increases or decreases, the regulator detects that and adjusts the base current to compensate, altering the transistor's resistance and permitting more or less current to go through it. The power lost in the resistance of the transistor is wasted and dissipated as heat.

Switching regulators chop the incoming current into fast pulses whose width can be varied. Those pulses are then applied to a capacitor, which charges up, converting the pulses back to smooth DC. By monitoring the output voltage, the regulator detects changes and alters the pulse width. The wider each pulse, the more current can charge the capacitor, raising the output. Narrower pulses lower it. While more complicated, this approach, called *pulse-width modulation* (PWM), supplies only the current required to keep the output voltage constant, without wasting the excess as heat. Thus, it is much more efficient.

PWM regulation is an inherent feature of switching power supplies, and some products use switching regulators internally as well. They are complicated, though, and also generate a fair amount of RF noise, so they aren't suitable for all uses.

Linear regulation, while wasteful, is still very common in low-current applications because the amount of power wasted is trivial. The linear approach is a lot simpler and cheaper, too, making it attractive.

While both types of regulators once took a bunch of components to implement, they can be had in chip form today, requiring just a few external parts to support their functions. The three-terminal linear voltage regulator, available in both fixed- and variable-voltage varieties, is the most common type you'll find. It looks like a transistor but is really an IC, with one terminal for input, one for ground and one for output. See Figure 7-15. Hang a capacitor or two on it and it's ready to rock. Some products have several regulators supplying separate sections of their circuitry.

Symbol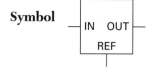

Markings

Voltage regulators are marked by part number, like transistors. Some standard markings tell you the voltage, which is handy. In particular, the widely used 7800/7900 series of linear regulators offers useful marking information. All regulators starting with *78* are positive regulators with negative grounds. A 7805 is a 5-volt regulator, a 7812 is 12 volts,

FIGURE 7-15 Voltage regulators

and so on. All parts starting with *79* are negative regulators with positive grounds. They use the same voltage numbering scheme.

Other regulators don't necessarily indicate their voltages in the part number, and you will have to look them up.

Uses

Voltage regulators provide stable voltage to entire devices or sections of them. In some applications, a regulator may feed a single stage or area of the circuitry. Some switching regulators can boost the voltage and regulate it at the same time. This is especially useful in battery-operated devices employing only one or two 1.5-volt cells but requiring a higher voltage.

What Kills Them

Pulling too much current through a regulator can overheat and destroy it. This is especially true with linear regulators. Voltage spikes can cause internal shorts, and random chip failures occur too. Switching regulators are prone to blowing their chopper transistors, just like switching power supplies.

Out-of-Circuit Testing

You can use the ohms scale to check for shorts, but that's about it. To evaluate a regulator properly, it needs to be in a circuit, receiving power.

Zener Diodes

Zener diodes are special diodes used in voltage regulators. In the forward direction, they conduct like normal diodes. In the reverse direction, they also act like regular diodes, blocking current. When the reverse voltage rises above a predetermined value set in manufacture, the zener breaks down in a nondestructive manner and conducts. This results in a constant voltage drop across the part, making it useful as a voltage reference. Zeners dissipate power as heat, so they are rated in watts for how much they can take before overheating and burning out.

Zeners look much like other diodes, but many have somewhat beefier cases and thicker leads to increase heat dissipation capability. See Figure 7-16.

Symbol

Markings

The band on a zener's case indicates the cathode, as with any diode. Since zeners are used for their reverse breakdown action, though, expect them to seem to face the wrong way in the circuit, with the band connected to positive. In fact, that's one way to help determine whether a diode you see on a circuit board is in fact a zener, and not just a normal diode.

Zeners are marked with part numbers, when they are marked at all. If the number begins with *1N47* and is followed by two more digits, that's a zener. Some have numbers like *5.1* or *9*, which would seem to suggest their breakdown voltages. This is *not* always the case! Look up the numbers to determine a zener's characteristics.

Uses

Zeners provide a stable voltage reference in many circuits, especially power supplies and regulators. Even switching regulators may use them for reference.

Zeners can be used as linear regulators by themselves when only a small current supply is required. A resistor will be used to limit the current going to the zener, and

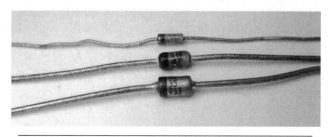

FIGURE 7-16   Zener diodes

the regulated voltage will appear at the cathode, while the anode goes to ground (in normal, negative-ground circuits).

What Kills Them

Putting too much current through a zener will exceed its wattage rating and overheat it, destroying the part. Over time, even zeners in proper service may fail from the cumulative effects of heating.

Out-of-Circuit Testing

Test zeners for basic diode operation using the diode test or ohms scales on your DMM. Zeners can short, but most fail open, or at least they appear to do so. In fact, they may short and pass so much current that they melt inside, quickly opening.

If the zener tests bad as a normal diode, it is bad. If it tests good, it may have lost its breakdown ability and still be bad, though. There's an easy way to tell, using your bench power supply. For this to work, the supply must be able to deliver a voltage at least a volt or so higher than the zener's expected breakdown voltage.

Take a 10 KΩ resistor (brown-black-orange) and put it in series with the zener's cathode, using clip leads. Connect your bench power supply with its + terminal to the other end of the resistor and its – terminal to the anode of the zener. Set your DMM to read DC volts and hook it across the zener. Turn the bench supply as low as it will go, and then switch it on. Increase the supply's voltage while watching the DMM. As the indicated voltage rises, it should hit the zener's breakdown point and the DMM's reading should stop rising, even though you continue to crank up the power supply.

If the voltage keeps going up past the zener's breakdown voltage, the part is bad. If it stops very near the rating, it's fine; standard zeners are not high-precision devices and may be off by a few fractions of a volt.

This test is also handy for characterizing unmarked zeners, as long as they are good. When you encounter a dead unmarked zener, determining what its breakdown voltage was supposed to be can be a real problem, unless you can find a schematic diagram of the product. Sometimes you can infer the breakdown voltage from other circuit clues. We'll get into that in Chapter 11.

There are many other kinds of less frequently used components. If you run across one that doesn't fit into any of the categories we've discussed, look up its part number to find out what the part does. An online search will usually turn up a data sheet describing the component in great detail.

Chapter 8

Roadmaps and Street Signs: Diagrams

Today's products can contain hundreds or even thousands of components. Even after you've considered a unit's failure history and pondered a preliminary diagnosis, there can still be lots to examine and test. How the heck do you find your way around what looks like a city on Mars?

There are three types of roadmaps to help you navigate the innards of an electronic device:

- **Block diagram**   This lays out the device by the function of each section and its basic interconnections. It does not show individual components or specific connection points. It's the most general, conceptual view, analogous to a map showing cities and route numbers for major highways between them, but not street-level detail. See Figure 8-1.
- **Schematic diagram**   This shows all the components and interconnections but does not indicate their purposes by specifying sections or overall structure. This is the street-level map. See Figure 8-2.
- **Pictorial diagram**   This uses drawings of the parts and shows their interconnections. The pictorials included in service manuals are really layout diagrams, detailing the placement of components as they exist on circuit boards and chassis. This is the drawing of landmarks and where to find them. See Figure 8-3.

All three diagrams work together to guide you to your destination. The block diagram helps you grasp the signal flow and interactions between circuit sections so you can see how they are supposed to work with each other. The schematic shows you individual components and stages so you can zero in on specific components you may want to scope or pull for testing. The pictorial helps you find the darned things!

If you can obtain only one style of diagram, get the schematic, because it offers the detail necessary for troubleshooting at the level of individual components. In years gone by, most products included a schematic, printed inside the case or in the instruction booklet. That became impractical as gadgets got more complex; there were

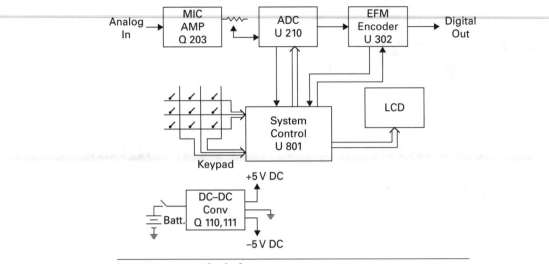

FIGURE 8-1 Block diagram

just too many parts to fit the diagram in such a small space. For awhile, manufacturers supplied fold-out schematics with their instruction booklets, but they finally began omitting the sheets as products no longer had incandescent lamps, snap-in fuses or any other parts the user could change. After all, only a service tech could really make use of diagrams anyway, so why spend the money to print them by the hundreds of thousands? Instead, service manuals were made available to repair shops, and the

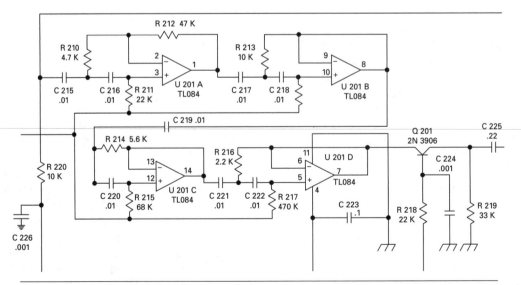

FIGURE 8-2 Section of a schematic diagram

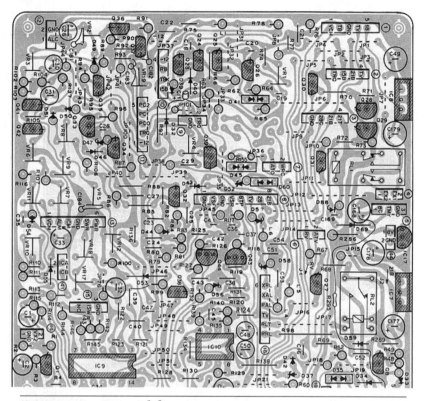

FIGURE 8-3   Pictorial diagram

end user, soon to be called the "consumer," was left high and dry, no longer privy to the products' insides. "No user-serviceable parts inside. Refer service to qualified personnel," replaced the diagrams. Is it any wonder they call today's gadgetry "consumer electronics"?

Service manuals were cheap and plentiful, and shops kept huge rows of filing cabinets bursting with them. In addition to manuals generated by the products' makers, the Howard W. Sams company produced its own comprehensive line of Photofact schematics for just about everything out there. If you couldn't get a schematic from Zenith, you could get a Sams easily enough for a buck or two.

As products got still more complex, manuals grew from a few pages to a few hundred, with large, fold-out schematics and very detailed, color pictorials. Producing these big books became quite expensive, so their prices skyrocketed. Shops continued to buy them—they had little choice—but no consumer would spend more for a manual than the product cost in the first place! Companies gradually reduced and finally abandoned the infrastructure for selling manuals to the public, and today's age of "use it, wear it out and toss it" was in full swing. Many manufacturers will no longer sell schematics or service manuals to consumers, thanks in part to fear of potential lawsuits by injured tinkerers. Some companies won't even sell manuals to service shops unless they're factory-authorized warranty service providers. And, believe it or not, some even refuse to provide diagrams to those facilities! Secretive computer makers, in particular, only let authorized servicers swap boards; the techs work on their machines for years without ever seeing a schematic of one.

Where does this leave you? Forget calling the major manufacturers; they won't sell you a service manual no matter how you plead. There are online sources, though, continuing to provide this vital information, at least for some products. Howard W. Sams continues as Sams Technical Publishing, at www.samswebsite.com. Their Photofact and Quickfact manuals aren't $1.50 anymore. As of this writing, most of them cost $20 and up. Still, for a tough case that has you going around in circles, it may well be worth the investment. Numerous other sites offer diagrams, some for free. Doing a Web search may turn something up, and it's always worth a try.

Hooked on Tronics

Reading a schematic is a bit like reading music: learning to name the notes is just the beginning. To really understand what's going on, you need to recognize the larger harmonic and rhythmic structures and how individual notes fit into and connect them. Identifying components on a schematic is a good start, but seeing how they form stages and sections, and how those work with each other, is vital to being able to find the ones that aren't properly performing their functions. The best techs have a good grasp of circuit fundamentals, but there's no need to be an engineer or a math whiz. It's far more useful to be familiar with the overall structure and with how basic circuit elements like transistors work.

In Chapter 7, we reviewed the component symbols for the most common parts. Along with those, schematics include symbols for other items. Here are some you're likely to encounter:

AC Voltage   The presence of AC voltage is indicated by a sine wave.

Antenna   The antenna symbol represents any type of antenna, even if it doesn't resemble the symbol.

Battery   In addition to the main batteries powering portable devices, small backup batteries may be found on circuit boards, either soldered or in holders. When wires connect batteries or battery holders, it is standard to use a red wire for positive and a black wire for negative.

Conductors, Joined   This is where two wires or circuit traces meet and connect.

Conductors, Not Joined   This is where two wires or circuit traces cross on the schematic (but not necessarily physically in the device) without connecting. Older schematics use the 3D-looking loop shown on the right. Newer diagrams don't, because today's products are so densely packed with interconnections that the loops become unwieldy and take up too much room on the schematic. Remember, if there's no dot where they cross, they are *not* connected! When a conductor meets another one without crossing, though, it is connected whether or not there's a dot.

(a)   (b)

Conductors, Merged   This shorthand description of multiple wires is widely used in digital gear, particularly when parallel data lines all go to the same chip. Instead of showing a separate conductor for each line, only one is shown, making complex schematics a little less cluttered and a bit easier to read.

DC Voltage   The lines represent what DC looks like on an oscilloscope, with the dotted line indicating ground.

Ground   There are four types. *Earth* (at left in the illustration) indicates a connection to the AC line's ground lug. *Chassis* (at right) means the connection goes to the unit's metal chassis or, lacking one, a common point on the circuit board. The earth symbol is often used in place of the chassis ground symbol—you'll see it in battery-operated gear that never gets connected to the AC line—but not the other way around.

Analog and digital ground symbols are used in devices having separate ground points for their analog and digital sections. This arrangement helps keep electrical noise generated by the digital system from intruding into sensitive analog circuits. Many CD, DVD and MP3 players have separate analog and digital grounds. In some products, the grounds meet, and only the length of a circuit board trace separates them. In others, analog and digital grounds remain separate, and connecting them externally will cause malfunction or undesired noises in the output signal.

Jack   There are many styles of jacks, so jack symbols are somewhat pictorial. Also, some jacks have internal switches that sense when a plug is inserted, and those will be shown too.

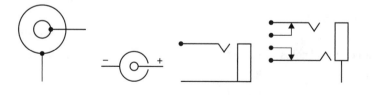

Speaker   The speaker symbol looks like a classic loudspeaker, but it can be used to indicate headphones as well. Sometimes a drawing of a headset will be shown instead.

Call Numbers

Each component will have a part number unique to that product's schematic. Some techs refer to it as the *call number*. That number is unrelated to the part number printed on the component, the one by which you can look up the part and learn its electrical characteristics. Instead, the call number is derived from the parts list found in the service manual.

Each call number begins with a letter specifying the type of component, followed by a few numbers. The first number tells you in what section the component resides, and the others are unique identifiers. Designations like R201, L17 and Q158 are call numbers.

The section number is arbitrary and varies from product to product, but parts with the same first number will live in the same neighborhood. For example, R201, Q213 and C205 will all be in the same area, but R461 and Q52 won't. And C206 is probably right next to C205, or at least not very far away. If the circuit board is labeled, it'll show those numbers next to each component.

Every part normally has its own call number, but multisection integrated circuits such as op-amps may be shown with each section as a separate device, even though they're really in one package. In Figure 8-2, the TL084 chip has four sections, each labeled A, B, C or D. On another schematic, the same component could be drawn as one rectangle resembling the shape of the real thing, with all four sections inside. In that case, each section's triangular op-amp symbol might be drawn inside the little box, but you can't count on that. Presenting the part as separate sections helps keep the signal flow clearer and is the preferred method.

Although each schematic has unique call numbers, the component type letters are somewhat standardized. Here are the letters in common use:

Component	Designator	Component	Designator
Capacitors	C	Relays	RL
Connectors	J or CN	Speakers	SP
Crystals and resonators	X or Y	Switches	S or SW
Diodes	D	Transformers	T
Fuses	F	Transistors	Q
Coils (inductors), but not transformers	L	Test points (places to put your probes)	TP
Integrated circuit chips	IC, U or Q	Voltage regulators	IC, U or Q
Resistors	R	Zener diodes	Z, ZD or D
Potentiometers	R or VR		

Because Q is the designator for transistors, it gets used for just about anything made from them, even if they are microscopic structures in an integrated circuit chip. Newer schematics, though, are more likely to differentiate, calling integrated circuits *U*, transistors *Q*, and voltage regulators *IC*.

Call numbers are handy even if you don't have a parts list, because they help you identify mystery components. Especially in this age of ultra-tiny, surface-mount parts, some look so similar that it's hard to guess what they are. If you see a call number starting with an *R*, you know the part is a resistor. An *L* tells you it's an inductor, and so on. Should you run into a part with a designator not covered in this book, a quick trip to the Internet will turn up its meaning.

The Good, the Not Bad, and the Miserable

Not all schematics are alike. There are good ones, even great ones. There are average ones. And there are the dreadful diagrams that are almost worse than none at all.

The Good

A good schematic is logically laid out, showing most stages with signal flow from left to right, with enough space between the stages to make the organization clear. It includes call numbers and part numbers, with resistor and capacitor values specified. A really good one may have arrows indicating signal flow through and between stages. A truly great diagram even has voltage readings and—it doesn't get better than this—snapshots or drawings of scope waveforms at various test points! If you're lucky enough to work with such a schematic, it'll greatly speed up your hunt. Touch a probe, compare what you see to the diagram, and either it looks the same or it doesn't. In real life, it's rarely as simple as that, but having those guideposts is a wonderful help.

The Not Bad

A merely okay diagram is clear, with a reasonable sense of organization. It has call numbers but probably no part numbers or parts values. Forget about signal flow arrows or waveforms. It's no GPS, but it's a serviceable roadmap. Everything is accurate and nothing is left out. Which brings us to the dark side, that malevolent maw of misleading misery, the incorrect schematic.

The Miserable

A really bad schematic may have reversed diode polarity, wiring errors, incorrect connection indications on conductors crossing each other, or omission of some parts, any of which can confuse the living heck out of you and send you off in the wrong direction. Switches show no indication of what they do in what position, and the drawing might not even be clear enough that you can read parts of it. Still, even a miserable schematic can be better than none, as long as you remember not to trust everything you're seeing. Occasional errors crop up in even the best diagrams, of course, but it's rare to find a truly rotten schematic from a major manufacturer. I've seen some doozies from off-brand companies, though.

Once Upon a Time...

To get started reading schematics, consider the organization of a book. It begins with letters that form words, which make sentences. Those are grouped into paragraphs and finally into chapters. Each paragraph links with the others to tell a story, and the chapters present a progression driving toward the finish. Some conditions are introduced at the start of the book and resolved at the end.

Electronic devices are organized much the same way. Components work together to form *stages*, each one feeding others, resulting in a signal flow proceeding from some starting point, such as a microphone, antenna or DVD, to some ending point, perhaps a display screen or a speaker. Each stage performs a function contributing to the overall processing. A group of stages involved in a particular part of the device's operation constitutes a *section* dedicated to a specific purpose.

With occasional exceptions, a schematic's signal flow in each stage proceeds from left to right. Signal flow *between* stages normally goes left to right as well. So, the most sensitive stages handling the weakest signals are usually on the left side of the page. If you're looking for an antenna or microphone input, look for it there. In a power supply, the AC line connection is probably on the left side, too, as it's considered the supply's input. Very complex schematics sometimes violate these conventions, simply because they run out of room on the page.

Look for power supply sections at the bottom of the page. Output sections, LCD screens and speakers should be on the right. Processing stages, such as the IF (intermediate frequency) stages of a radio receiver or the microprocessor in an MP3 player, will be in the middle.

When reading a schematic, keep your eye on the story and its central characters. Not all players are equally important. The plot is driven by the active elements like transistors and ICs, since they do most of the work. Crystals and resonators generate signals, so they're crucial characters, without which the story never gets moving. After those, look for coupling elements such as transformers, capacitors and resistors linking one stage to the next. They move the plot forward, because signals will be flowing through them on the way to subsequent stages. The other resistors, capacitors and coils set the voltages, currents and various conditions the big shots need to do their jobs. Those subplots are necessary to the overall story but not central to its theme. Try not to let them distract you from the primary action. Let's look at the schematics of a few circuits and how to interpret them.

Amplifier Stage

For our first adventure, we'll look at a single stage. Let's examine every part in it, what it does and what would happen if it malfunctioned.

This one is an inverting amplifier, typical of what you might find in just about any product. See Figure 8-4. *Inverting* means that the output signal rises as the input signal falls, and vice versa, producing a replica that's upside down. Sometimes inversion is necessary to the circuit's operation, while other times it's just an irrelevant consequence

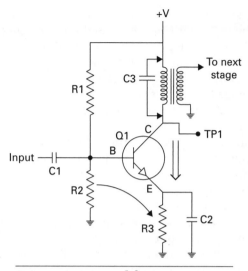

FIGURE 8-4   Amplifier stage

of getting voltage gain from the stage. Either way, you'll see lots of inverting amplifier stages.

The stage has four major points where things go in and out. At the top, power is applied through the transformer, reaching the collector of the transistor. Signal input is on the left and goes to the base of the transistor, Q1, via C1. The output is at Q1's collector, which is why TP1, the test point, is there. Finally, R2, R3 and C2 go to ground. Ground is as important an input/output point as the others; without ground, you've got a paperweight.

To be a sharp troubleshooter, you need to understand how the circuitry is supposed to function. So, let's see how this thing works, starting at the top. Power goes through the winding of the transformer and reaches the collector of the transistor, Q1. It's the *active element*, so it's the central character. When the transistor is turned off, no current passes through the transformer, because the lower end of its winding sees no connection to ground. When the transistor is turned on, the path between collector and emitter connects, effectively grounding the transformer's winding and completing the circuit, pulling current through the winding.

It's not an all-or-nothing proposition, though. Remember, transistors act like variable resistors (potentiometers), except that signals, rather than your fingers, turn them up and down. Here, the signal is applied to Q1's base through C1. C1 couples only the AC component of the incoming signal to Q1, preventing any DC in the signal from reaching the transistor, and also preventing any DC *from* Q1 getting back into the previous stage.

Where R1 and R2 meet establishes a voltage somewhere between the power supply's value and ground, to bias the transistor's base, or put a little DC on it, keeping it turned on through whatever portion of the incoming signal's waveform the amplifier is intended to amplify. As with most amplifiers, we want the whole waveform, so the transistor has to be biased with enough positive current that when the incoming signal goes negative, the transistor's base never gets below about 0.6 volts, which is the cutoff point for a standard bipolar silicon transistor. The bias current has to flow through R2, so its value determines how much is available to the transistor's base; the higher R2's resistance, the less bias current there will be.

The transistor passes current from collector to emitter in proportion to how much current passes from base to emitter, as shown by the arrows. The ratio of the two currents determines how much the transistor can amplify a signal. It is an inherent part of the component's design and is given in its specifications. As the incoming signal wiggles up and down, the base current varies with it, causing the transistor to pull a proportionally larger current through the transformer on its way to R3 and finally ground. That forming of the power supply's DC into an enlarged replica of the signal is called *gain* and is the essence of amplification and the foundation of all modern electronics, analog or digital.

Ah, R3 and C2. What're they there for? R3 limits the total current through the circuit; without it, the transistor would attempt to pull the power supply's entire current capability to ground, dragging down the supply and probably blowing the transistor or the transformer. R3 limits the total base current as well, because it's in series with that path too. Thus, it sets a limit to how much signal current the previous stage has to supply.

C2 is a little trickier to explain. Transistors, having adjacent regions of semiconductor material in them that are not at the same voltage at the same time, also behave like capacitors. They store some charge and take a little time to discharge. The presence of R3 slows that down because the discharge has to reach ground through its resistance. This forms a *time constant*, which is a fancy way of saying that there's an upper limit to how fast the transistor can get rid of its charge. The bigger the value of R3, the longer the discharge process takes, and the slower the transistor can react to incoming signal changes. When the frequency of the incoming signal is faster than the time constant, the transistor's residual charge fills in as base current drops, acting like any filter capacitor and smoothing out the waveform. As a result, the transistor can't react quickly enough to respond and amplify the signal. Thus, the upper speed limit, or *frequency response*, of the amplifier drops off.

C2 allows the rapidly changing parts of the signal to reach ground with less resistance, discharging the transistor faster. The apparent resistance of a capacitor drops with increasing frequency, because it never gets the chance to charge fully and oppose the incoming current. So, C2 compensates for the transistor's capacitance, giving it a lower-resistance path to ground with increasing signal frequency. The result is to restore the lost high-frequency response of the amplifier without also increasing the low-frequency response.

T1, the transformer, plays a crucial part in the amplifier's operation. As current is pulled through its *primary* coil, it generates a magnetic field that impinges on, or cuts across, its *secondary* coil, the one on the right. As the current in the primary gets stronger and weaker in step with the signal, the changing magnetic field generates a current in the secondary that makes its voltage rise and fall in step. That changing voltage couples the new, amplified signal to the next stage. There are other ways to couple a signal, without a transformer, but using one has advantages in some kinds of circuits, especially those employing the transformer as a *tuned circuit* resonating at a specific frequency. Radio receivers use lots of tuned amplifiers of this sort to pick out the selected signal from all the others hitting the antenna. A capacitor across the transformer, shown in Figure 8-4 by the optional C3, is a dead giveaway that an amplifier is tuned. C3 could be on the other side of the transformer, too, and still have the same effect in tuning it to a desired frequency.

Thar She Blows

Now that we've explored how the amplifier stage is supposed to work, let's see what the effects of malfunctioning parts would be. Again, starting at the top, how would a bad transformer affect the performance of the circuit?

If the transformer were open, no current would pass through its primary, so the collector of the transistor would read 0 volts. Transformers in small-signal circuits don't pass much current, so an open winding is unlikely, but it can happen. By the way, to estimate how much total current the winding might have to handle, just divide the power supply voltage by the value of R3, a la Ohm's Law. This pretends that the transistor and the transformer's winding have no resistance, so the real value will be a bit less, but at least you'll know the approximate upper limit.

If the transformer had a short, its resistance would decrease and it'd be harder for the transistor to pull its lower winding connection toward ground, so the voltage at the transistor's collector would be close to that of the power supply, with little signal variation. Depending on the total current through the circuit (limited by R3, as described above), the transistor might get hot or even be blown.

If R1 were open, there'd be no bias current going to the transistor's base, so the transistor would be turned off. Its collector would be at the same voltage as the power supply, and its emitter would be at ground potential, 0 volts. The signal itself might have enough current to turn the transistor on a little bit during the positive half of its waveform, resulting in a weak, distorted mess of negative-going signal excursions appearing at its collector. (Remember, the amplifier inverts.)

Shorted resistors are pretty much unheard of, but, for the sake of this thought experiment, we'll consider what would happen if one did short. If R1 shorted, the base would be biased to the power supply's full voltage. The transistor would be fully turned on no matter what the incoming signal did. The collector would read somewhere close to 0 volts, and the transistor might be hot, as might be R3.

If R2 were open, the bias would be too high, with much the same result. The signal's influence might result in some output, though only on the negative-going half of its waveform. Those negative incoming peaks would cause a rise at the output, thanks to the inversion. As before, you'd get a weak, distorted mess, but the collector's DC level would be low, not high.

If R2 were shorted, the base would be pulled down to 0 volts, resulting in the same no-bias conditions you'd get if R1 were open, except that the signal could not produce any output at all, because it would be shorted to ground as well.

If C1 were open, no signal would get to the transistor, but the DC voltages on Q1 would look just fine.

If C1 were shorted, the base bias would be influenced by the previous stage, resulting in unpredictable behavior. If the bias were pulled toward ground, Q1's collector would rise, clipping off the top of the output signal if it got too high. If the bias went high because of voltage being fed in from the previous stage, the transistor would turn on too hard, the collector would be low, and the signal would be cut off toward the bottom.

A shorted or leaky input coupling capacitor is something you might actually run into, particularly when it's an electrolytic or tantalum cap. With a leaky one, it can be maddening to try to deduce why the stage behaves so oddly. If you disconnect the input cap and the DC voltages on the transistor change, the cap is letting some DC pass through it. It's leaky.

Now to the heart of the stage: the transistor. This is the component most likely to cause trouble. Transistors can fail in numerous ways. They can open from emitter to collector, usually as a result of overcurrent. They can also short that way, and often do. A short or an open can occur from base to emitter as well. I've seen transistors with all three leads shorted together like one big piece of wire!

Transistors can be leaky, too, allowing current to pass when it should be cut off, or even to move backward through their junctions. A very small amount of reverse leakage is normal, actually, but it's not enough to affect the circuit's operation. A leaky transistor allows much more reverse current, and that'll produce all kinds of unpredictable effects. The little monsters can also become thermal, changing their gain and leakage characteristics as they warm up.

An open between any two junctions will result in no output. TP1 will be at the power supply voltage, because the transistor will not pass any current toward ground. Even if the collector-emitter junction is fine, an open base junction will prevent the transistor's being turned on.

A shorted collector-emitter junction, which is quite common, will appear as if the transistor were turned on all the time, all the way. The collector voltage will be at or very near zero. A shorted base-emitter junction, also common, will pull the base bias and incoming signal down toward ground through R3, and the transistor will not turn on. So, the collector voltage will not pull down and will be at the power supply voltage. A short from collector to base will turn the transistor on, having the opposite effect.

A quick-and-dirty way to hunt for transistor shorts is to check the DC levels on all three leads. If any two are exactly the same, the part may very well be shorted. If they're even a little bit different, a short is far less likely.

If a transistor has leakage, it can act strangely. If collector current flows into the base, for instance, it can overbias the part. Depending on how much current leaks, the transistor may still work to some degree. If a stage acts wonky but everything seems to measure okay, and especially if the behavior changes with temperature, leakage is likely.

Switching Power Supply

Switching supplies range from moderately complex to ridiculously so. Rarely will you find one that looks especially simple. If you think I'm kidding, crack open a modern AC adapter. Even those inexpensive little wall warts are stuffed with chips, diodes, transistors and regulators, along with a smallish transformer and the usual plethora of resistors and capacitors.

Switchers producing only one output voltage are less dense, with a straightforward arrangement for regulating it. More elaborate circuits can have multiple output voltages, overcurrent sensing, and other fail-safe protection measures adding to their parts counts.

At their core, though, switchers are not that complicated. Their basic operation is pretty much the same, regardless of the frills. So, let's strip away the doo-dads and look at what makes these omnipresent beasts purr.

Take a look at Figure 8-5 for a simplified schematic of the sort of power supply you're likely to encounter in many modern products, from LCD monitors to big-screen TVs and audio gear. How do we know it's a power supply? The presence of the AC line input at the far left is a dead giveaway. Is it a linear supply or a switcher? Notice that the AC line goes directly to a bridge rectifier and then to a transistor, before reaching the transformer. That's the classic switcher design: the AC gets changed to DC, stored in a big electrolytic capacitor, and then chopped by a transistor which feeds the transformer. So, this definitely is a switcher. In a linear supply, the AC would go straight to the transformer; rectification, filtering and regulation would be done on the other side.

That big transistor below the transformer is the *chopper*. It switches on and off at a high frequency, pulling current through the transformer to generate a pulse of magnetism each time it turns on. The chopper is connected to the AC line, along with everything else on that side of the transformer.

The transistor is driven by pulses from the pulse-width modulator. That chip keeps an eye on the supply's output voltage and adjusts the *duty cycle*, or on-off ratio, of the pulses it feeds to the transistor's gate. The wider the pulses, the more energy will flow across the transformer, and the more power will fill up the output capacitor, keeping the voltage from sagging when the supply's load increases. If the load decreases and the output voltage starts to rise, the chip notices that and narrows the chopper's pulses,

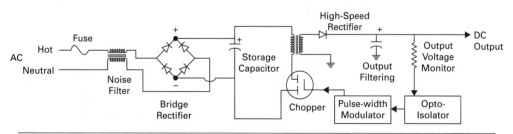

FIGURE 8-5   Switching power supply

bringing the output back down to its correct value. That regulation effect keeps the output voltage steady as the circuit being powered varies its demand for current.

Between the output stage and the pulse-width modulator is an opto isolator, which is nothing more than an LED and a light-sensitive transistor in one case. It's there to pass information about the output voltage back to the chip without forming an electrical connection between the two. Lack of a connection keeps the output side of the supply (everything to the right of the transformer) isolated from the AC line, and thus safe. Pretty simple story, isn't it?

Pop Goes the Switcher

Let's look at what would happen if the major components failed. Starting with input from the AC line, our first major component is the bridge rectifier. It could fail in several ways, but the most common problem is an open circuit in one of the four diodes. That'll result in half of the AC waveform's not getting transferred to the output of the bridge. With no capacitor to smooth things over, it would look like Figure 8-6.

Since a capacitor is storing the charge, you'll see a much-lower-than-normal voltage at its positive terminal, along with a droop where the missing segment should be replenishing it. See Figure 8-7. The chopper may still run in this condition, but it probably won't.

One of the diodes could short, resulting in reverse-polarity voltage getting where it shouldn't. In that case, expect the supply's fuse to be blown.

The most common failure in a switching supply is a bad chopper transistor. It operates at high voltages and takes a lot of stress. If it's shorted, the fuse will be popped and the supply will be dead as a doornail. If the transistor is open, the supply will still appear dead, of course, but the fuse will be good and the big capacitor will have a full charge of a few hundred volts on it.

If the PWM IC is dead, there will be no pulses at the base or gate of the chopper. The IC could appear bad, though, due to other factors. First, it needs some voltage to

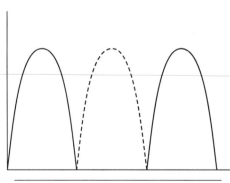

FIGURE 8-6 Missing rectified AC segment

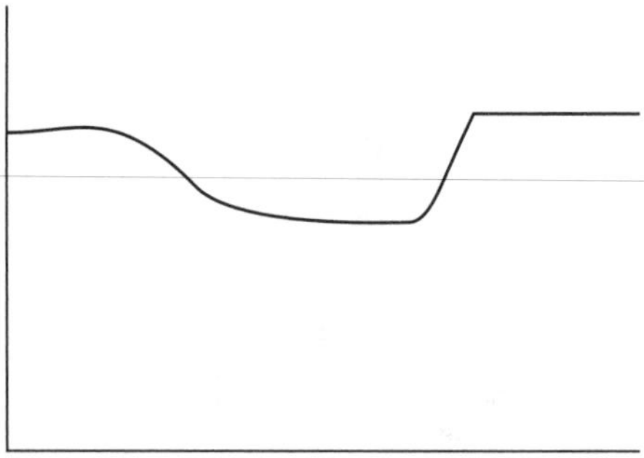

FIGURE 8-7 Droop due to missing segment

run, even before the chopper starts, so there could be a bad diode, zener or small cap in its standby voltage supply. Also, if the output voltage of the entire switching supply goes abnormally high, the chip will sense it and shut down. Typically, though, it will try to restart every second or so, resulting in a chirping noise from the transformer. Most switchers will also do that if the load they're driving pulls too much current, dragging the voltage down below what the supply can replenish.

Some switching supplies employ a *crowbar* circuit intended to blow the fuse if the output voltage goes too high. Crowbars usually employ a *silicon-controlled rectifier*, or SCR. This component resembles a transistor, with three leads, but is really a rectifier with a control gate. Its symbol looks like a diode with a third lead sticking into the junction. As with a transistor, an SCR's control gate turns it on, but that's where the similarity ends. Once tripped, the SCR stays on until the AC waveform changes polarity, regardless of the control input. Also, the gate can only turn it on, not off.

SCRs make great crowbars. Placed directly across the AC line, but just *after* the fuse, the SCR normally doesn't conduct, so it has no effect on the circuit. If the output voltage goes abnormally high, a detector circuit trips and sends voltage to the SCR's control gate, turning it on. The short across the line blows the fuse and stops the supply, protecting whatever gear it's powering, along with the supply itself. If you run into a switching supply with a blown fuse, but the chopper and bridge are good, it's reasonable to suspect that the crowbar tripped, and there's some problem with the voltage regulation system. Open zeners in the regulator circuits often cause this condition.

Push-Pull Audio Amplifier

Let's try another example. Figure 8-8 shows a slightly simplified channel of a typical audio amplifier. The design has the somewhat humorous but also descriptive name "push-pull" because it splits the incoming audio waveform into two halves, separately processing the positive and negative portions of the signal. One half of the amplifier pushes the speaker cone outward and the other half pulls it back in. It's how most modern audio amplifiers are built.

The example shown in Figure 8-8 is a true bipolar circuit, powered by positive and negative voltages with respect to ground, shown by + V and –V. The only components connected to ground are the filter capacitors, C5 and C6, on the power supply *rails*, and the speaker.

Some similar amplifiers use only a single polarity. That works fine, but it means that there will be a DC offset at the output of half the supply voltage, so that the signal can swing equally up or down before hitting the limits of the rail or ground. Such a unipolar design will have an output capacitor to block that DC offset from reaching the speaker and keeping its cone pushed halfway out (not to mention wasting a lot of power and heating up the speaker's voice coil).

Working from left to right, as usual, we see the input stage, which amplifies the incoming signal enough to drive the next stage, consisting of Q2 and Q3. The opposite polarities (NPN and PNP) of those transistors mean that opposite halves of the signal will turn them on. The signal is coupled to them by C3, which blocks any DC

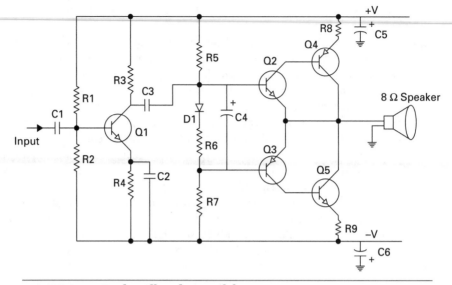

FIGURE 8-8 Push-pull audio amplifier

component from Q1 from influencing their biasing. The bias network of R5, D1, R6 and R7 keeps the transistors turned on just slightly, so there's no dead spot to cause *crossover distortion* when the input signal's waveform is less than ± 0.6 volts or so. C4 couples the signal to both transistors. Their outputs drive Q4 and Q5, which provide enough current gain to move a speaker cone. In a real amplifier, a little bit of the output would be fed back through a few resistors and capacitors to the input stage in a *negative feedback loop*, to correct for distortion introduced by the imperfect nature of the amplifying elements (transistors) by reintroducing the same distortion upside down, cancelling it out. We're omitting those parts here to keep things clear.

Sounds Like a Problem

Let's look at how malfunctions in each stage would affect the amplifier's behavior. If Q1 or its surrounding components broke down, no signal (or perhaps a very distorted signal) would emerge from C3. Because C3 blocks DC, the badly skewed voltages at the input stage would not affect the rest of the circuit, so further transistors would not be damaged by being turned on too hard and pulling too much current.

In a *direct-coupled* circuit, though, the input stage would be carefully designed to have no significant DC offset, and there'd be no C3. Should the input stage of a direct-coupled amplifier malfunction and generate a lot of DC offset, as may happen with a bad transistor, that offset could wreak havoc on the rest of the transistors, possibly blowing all of them.

If D1 opened, the top half of the amp would be turned on very hard by the bias provided by R5. Q2's base would go very positive, turning it all the way on. That would pull the base of Q4, the PNP transistor, down toward ground through the speaker, turning that transistor all the way on as well. Both transistors would get quite hot and might be destroyed.

Meanwhile, R7 would pull the base of Q3 very negative, which would turn that PNP transistor all the way on as well. That would turn on Q5, too, and funerals for Q3 and Q5 would likely be in order.

Normally, a push-pull amplifier has one half on at a time, with the other half conducting only slightly until the signal's polarity flips, reversing the process. Power supply current passes through the output transistors, one at a time, to the speaker and then to ground. In this case, both halves would be turned on at the same time, effectively shorting the +V and –V lines through the output transistors. Yikes! You can imagine the results. Smoke, burned emitter resistors (R8 and R9), blown transistors, an unholy mess...all from one bad diode.

If D1 shorted, though, the results would be different. The bias would become unstable and the amp's DC offset would swing around with the signal and distort it badly, but the transistors would probably survive because the bias wouldn't be so far out of whack that it'd turn them all the way on.

If C4 opened, the top half of the amp would work but the bottom would get no signal, so severe distortion would occur, with only one-half of the waveform present at the output. If C4 shorted or got leaky, the result would be similar to what you'd see if D1 shorted: the bias would get wonky, the amp would distort, but parts probably wouldn't be damaged.

An open in Q2 or Q3 would turn off the corresponding half of the amplifier, with loss of one half of the signal waveform. A short in one of those transistors would be a much more serious matter.

Q2 and Q3 are referred to as *driver transistors*, because they drive the output transistors, Q4 and Q5. At this point, the circuit is direct-coupled, so a short in a driver will turn its output transistor fully on, probably blowing it. At the very least, there will be a lot of DC at the speaker terminal, and the speaker may also be blown from all the current passing through it.

Mega Maps

Highly complex schematics can be tough to follow, with all kinds of confusing signal and power lines running every which way. Especially with such schematics, a block diagram can be incredibly helpful. Devices with multiple boards may have many connectors and cables shuttling signals back and forth. The connectors are valuable focal points. If you're not sure what goes where, trace the schematic back from a connector to see if you're in the right place. Still not sure? Grab the block diagram and see if that area goes where you think it does.

Another great place to find a signpost on a big schematic is an input or output point. Jack and speaker symbols stand out because there are so few of them on any given diagram. Trace the lines back to find what's feeding them.

Transformers also jump off the page. Whichever symbol you choose, you can train yourself to scan a schematic and find it, disregarding the others. It's a lot harder to do that with resistors and capacitors, of course, because there are so many of them.

Give It a Try

If you have any schematics, now's a good time to pull them out and practice your reading. If you don't have any, go on the Internet and search for some. Radios, TVs, and CD and DVD players all make for good reading material. See what you can identify. Try to find the following stages and sections.

Radios

Look for the *front end*, which is the first section accepting input from the antenna. In an analog radio, you'll see tuned circuits with variable capacitors. See if you can follow the input signal up to the mixer, which is where it gets mixed with the *local oscillator*.

Digital sets don't have variable caps. Instead, look for a big chip and a bunch of surrounding parts forming a *frequency synthesizer*. How do you find that? It'll be connected to an LCD and probably to a keypad, too.

Once you've found these stages, follow the signal through the mixer. The mixer may be a chip, or it may be four diodes in a ring configuration that looks a bit like a bridge rectifier, except that the diodes are facing different ways. Some mixers use a dual-gate MOSFET transistor, with the incoming signal fed to one gate and the local oscillator fed to the other. The mixer's output goes to the IF, or *intermediate frequency*, stages. Look for transformer coupling between stages. Most modern radios, analog or digital, convert the frequency twice (and sometimes even more), so you should see a fixed-frequency oscillator feeding another mixer stage, followed by more IF stages.

From the IF section, the signal gets *demodulated*, or detected. This is the process of extracting the information—audio, video or data—that was originally impressed on the signal at the transmitter. AM detection may involve nothing more than a diode and a capacitor. FM is a bit more complex, and data can involve all sorts of decoding circuitry. Data detection used to require a lot of chips, but these days a small microprocessor or a DSP (digital signal processor) chip may do the work. DSPs get used for enhancing voice signals, too, especially in modern communications receivers and transceivers.

CD and DVD Players

Despite the low prices of disc players, getting the data off an optical disc is not a simple task. It involves three *servo* systems working together to find the microscopic tracks, follow them as they pass by, and keep constant the rate at which their data is read out. The laser head must properly track the absurdly tiny groove, even though normal eccentricities in the geometry of the discs are many, many times the size of the grooves themselves. This is a three-dimensional problem, with the wobbling distance between the head's lens and the disc surface requiring a dynamic focusing servo to keep the beam size at the point it meets the track small enough to grab just one bit of data at a time.

Unlike analog records, CDs and DVDs use *constant linear velocity* (CLV) to pack the data in with maximum space efficiency. With CLV, the speed at which the laser head sees the track go by is constant, regardless of whether the distance around the track is short, as it is near the center of the disc, or long, as it is near the outer edge. So, the disc must spin faster at the start and gradually slow down as the disc plays, reaching minimum rotational speed as the head plays the longest tracks, nearest the outside. Accomplishing this automatic speed control requires yet another servo to keep the disc spinning at exactly the required rate.

Take a look at a disc player's schematic. Can you find the laser head? Notice that it has more than one photodetector (light-sensitive transistor). Three are used to keep the laser beam centered on the track. Look also for the servo coils used to float the lens and make it dance in step with disc wobbles.

Trace back from the head and see if you can locate the head *preamp*, which boosts the weak signals from the photodetectors. It should be a chip of medium density and will have a few test points at or very near some of its pins.

Trace back from the servo coils and see if you can find the focus servo section and the tracking servo, too.

Find the sled motor, which moves the head across the disc as it plays, and the circuit driving it. Most motor driver circuits have transistors between the chips controlling the motion and the motor because the chips can't supply enough current to run the motor directly. Since the sled motor has to run in either direction, look for what's called an "H bridge" configuration of the driving transistors, in which the connection to them is in the middle, with each wire going to the motor coming from where two transistors meet. It looks like the letter *H*, hence the name. Neither wire goes to ground, so the controlling circuitry can flip the polarity to the motor at will, reversing its direction. Some H bridges are implemented on a single chip, but many are still made from separate transistors. See Figure 8-9.

Now search for the disc motor, which spins the disc. Its driver circuitry will look similar to the sled motor's circuit, but not identical. It'll probably have a transistor as well, but it has to spin only one way, so no H bridge is required.

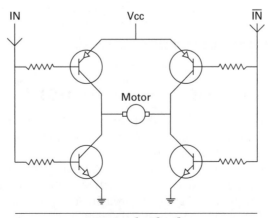

FIGURE 8-9   Typical H bridge

Keep Reading

You'll find lots of schematics on the Internet. Get a few and practice reading them, focusing on signal flow and organization. Try to deduce which components are generating or passing signals, and which are support systems for the central players. Look for coupling components, filter and bypass capacitors, power supply sections, digital control systems, voltage regulators, oscillators, and so on. After awhile, reading a schematic will be as familiar as reading a book. You'll be able to take one look and recognize the sections and stages.

But I Ain't Got One!

As important and useful as a schematic is, you will find yourself working on many devices without one, simply because you can't get it or it costs more than you want to pay. Without the roadmap, how do you find your way around?

It's a lot tougher, but if you keep the overall circuit functions in mind, you can find the major sections and determine whether they're working properly. When you get to a suspicious stage, try drawing your own mini-schematic by tracing the connections of the active element and its surrounding components on the board. Sometimes seeing it in front of you will illuminate the concept of the design and lead you to good troubleshooting ideas.

Once you locate the malfunctioning section, your understanding of stages and signal flow will help lead you to the problem. At least, that's the way we want things to go. Sometimes it does, sometimes it doesn't. In very complex devices, it's easy to get lost and go around in circles when you're flying blind, without a diagram. Luckily, most malfunctions are due to power supply problems, bad connections and faulty output stages, which are relatively easy to track down, even without a schematic. When complex signal processing stages don't work right, you may not be able to determine why, especially in modern devices with hundreds of parts the size of grains of salt. Even in those situations, though, you can learn a lot by tracing back from input and output jacks and poking a scope probe on what look like input and output points of successive stages. You just might find the spot where the signal disappears and zero in on the bad component.

Your Wish Is Not My Command

Let's take a case and work through it without a diagram. This will be a nice little LCD TV that refuses to respond to its remote control. It functions fine with the front-panel buttons, but the remote does nothing.

Is the remote working? How can we tell? If only we could see infrared light! Got a camcorder or a digital camera? Those can see infrared. Even though they have filters to block it, some IR light gets through. Point the remote at the lens and hit one of the buttons while looking at the camera's display screen. If you see a flashing light, the remote is working. Could it work but not be sending the right codes? Yes, theoretically, but I've never seen it happen, except in the case of a universal remote set up to operate the wrong device. That's user error, not a repair problem. If the remote's IR LED lights up when you press a button and stops when you let go, you can assume it is working properly.

In this case, the remote works. So, why can't the TV see it? Something in its remote receiver circuitry is out, and we're going to hunt that problem down. Naturally, this newer product has no available schematic, so we're on our own.

The first thing to find is the photodetector that picks up the remote's signal. Most products use a prefab remote receiver module containing a photodetector and a preamp. The module is usually in a little metal box (see Figure 8-10), though in small, battery-op gadgets it's more likely to be a bare plastic part. It sits just behind the front panel of the unit so the photodetector can see out through the panel's plastic bezel. Yup, there it is. It's mounted to the circuit board and connected by three lines. Let's stick our scope probe on them and see what's there.

First things first: we have to connect the scope's ground lead to circuit ground and then power up the TV. To begin, let's set the scope for DC coupling, 1 volt per div, and put the trace at the bottom of the screen, since we expect positive voltages in this negative ground TV. (We did verify that the negative power supply line went to circuit ground, right? Of course we did.) Hmm...one of the module's three lines seems to have nothing on it at all. Why would a connection have no signal? Could finding the fault be *this* easy? Nah! The line goes to a big area of copper on the board. It's ground! Okay, one down. The second line shows a steady voltage that doesn't vary when the remote is pointed at it and a button is pressed. That one must be the power

IR Module

FIGURE 8-10   Remote control receiver module

supply voltage running the module. The third line has what looks like a little noise on it. When a remote button is pushed, that line becomes an irregular pattern of pulses at just under 5 volts peak to peak. Aha! That's the code the remote is flashing. We've found the module's output line.

The good news: we've proved the receiver module is okay. If it weren't, the output line wouldn't do a darned thing when the remote sent its optical signal. It's showing the transmitted code, so it's working. The bad news: we still have no idea why the TV won't respond to it.

So, we follow the output line and see where it goes. It appears to terminate at a diode that feeds a transistor. The diode's purpose isn't obvious, but there it is, so we'd better check it. We scope it to see if the signal from the module is getting there. It is. At the other end of the diode, the signal is smaller but it's still there. Because the signal is smaller, we can figure that the diode is not shorted. And, because it's there at all, the diode is not open. It could be leaky, but most likely it's fine. Time to focus on the transistor.

One of the transistor's three leads, presumably the base or gate, is connected to the diode. Sure enough, the remote's signal is there. The other two leads are at 0 volts, though! No DC, and neither shows any activity when the remote is activated. No wonder the TV never sees the remote's signal. It's a dead end. Blown transistor! At least, that's what a novice would think. In truth, we don't have enough information to draw that conclusion yet, so we look further.

One lead goes to circuit ground, so we wouldn't expect anything there. The other should have some voltage, though, right? Tracing from that one, we come to a couple of resistors, and it isn't clear where they go. Since the transistor needs power to function, one of those resistors must go to a power supply feed point. And since the transistor's output signal has to feed some other part of the circuit that knows what to do with it, the other resistor must be coupling it to another stage. But which one does what?

Putting the scope probe on the opposite end of each resistor, we find that neither has anything on it there either. To use the technical term, *deadus doornailus*. If the transistor were shorted, that would pull its output line to ground, but not the other end of the resistor feeding power to it. That's the whole point of having a resistor there: to limit the current and avoid pulling the whole supply down when the transistor turns on and connects the line to ground. So, the resistor's far end, fed from the power supply, should still have voltage on it no matter what the transistor does. But the voltage isn't there. This suggests the problem lies elsewhere, and the transistor is probably not the culprit.

Tracing each resistor, we see that one goes directly to a huge chip. The other goes someplace far away we can't easily find. The one going to the chip is most likely the transistor's output, feeding the remote's signal to the microprocessor for decoding. The other one, then, has to lead to the missing voltage. So, we follow it as it snakes along. Eventually we find it leading to a little jumper wire, the other end of which connects to a fairly large land with lots of other parts and traces going to the same place. That should be the power supply feed point. Poke ye olde probe, and there's five lovely volts. Yee hah, it works! Oh, wait a minute, why isn't it getting to the resistor?

We turn the set off and watch the voltage on the feed point die away to zero. Out comes the DMM, set to the ohms scale. That's odd; there seems to be no connection from the feed point to the resistor. A check of our tracing confirms we haven't made a mistake. Nope, we're in the right place. We flip the board and take a good look at that jumper wire's solder joints. One of them looks cracked! We touch it up with the iron and recheck the line for continuity. Now there's a connection! You can guess the rest. We power up the set and it works, remote and all!

That is a *very* typical repair case. And, as you've seen, it can be solved without a diagram, by understanding basic circuit function and applying a little logic. Had there been voltage at the power supply side of the resistor but not at the transistor it was feeding, yanking the transistor and checking it for shorts would have been the next step, and it's a pretty safe bet the part would indeed have been shorted from collector to emitter, pulling the applied voltage to ground. The lack of voltage at the resistor's far end was the critical clue in acquitting the transistor, tracking down the real culprit and solving the case. Verdict: time served, and free to go!

Chapter 9

Entering Without Breaking: Getting Inside

Once upon a time, getting at the circuitry of a consumer product was a piece of cake. Remove a few screws, pop off the back, and there it was. Providing access to the inner workings was a tradition begun in the vacuum tube days, when the unit's owner needed to get inside on a frequent basis to change tubes, lamps and fuses. After semiconductor technology replaced the troublesome tubes with considerably more reliable transistors, there was still the expectation that the buyer might have a legitimate need to reach the circuit board. Early solid-state products even put the transistors in sockets! As semiconductors got sturdier, the sockets went away, which was good since their flaky connections caused more failures than did the transistors themselves. There were still lamps and fuses to contend with, though, and access was expected and easy.

That was before today's age of complex, ultra-miniaturized circuits, complete lack of user-serviceable parts (mostly thanks to the LEDs that replaced lamps), and lawsuits. These days, no manufacturer wants you anywhere near the stuff under the hood.

Consequently, many equipment cases are deliberately sealed, or at least made pretty tough to open. Most AC adapters are ultrasonically welded together and have to be cracked open; even the expensive ones are considered non-repairable items by their makers. Lots of laptop computers, video projectors, MP3 players and TVs sport hidden snaps, so they won't pop apart even after you remove the screws. See Figure 9-1. And anyone who's ever tried to take apart a certain American computer company's sleekly designed products knows the meaning of frustration; they're clamped together seamlessly and tightly in a clearly deliberate attempt to keep you out.

So how do you open up these crazy things? Sometimes, others have suffered before you and have posted step-by-step disassembly instructions on their Web sites, with clear photos of the whole process. If you can't fathom how to get something apart, it pays to do a Web search on "disassemble _____," with the blank filled in by the name or model number of your product. Some companies selling parts post these

FIGURE 9-1   Hidden plastic snap

helpful tutorials, facilitating your getting to the point of having a reason to order their wares.

If you can't find disassembly instructions, you'll have to wing it. With or without help, this is the stage at which you have the most opportunity to wreck your device! Without X-ray vision, you can't know if that tiny screwdriver you're using to pry the case halves apart or unhook the hidden snaps you're not even sure are there might be ripping a tiny component off the circuit board or causing some other drastic damage. You also have no idea whether a ribbon cable may join the two halves and be torn when they suddenly separate. And nothing is more frustrating than thinking you're about to repair something and destroying it instead, before you ever get a chance to look for the original problem.

Despite all this gloom and doom, you can get into nearly any product successfully and safely if you're careful and take your time. Disassembly is not a trivial process; expect it to take a significant portion of the total repair time, at least with the smallest, most complex devices. There are some tricks to the endeavor, and we're going to explore them now. But first, some rules:

- **Rule number one**   *Always* disconnect power before taking something apart. This is true with battery-operated products as well as AC-powered ones. Even if there's no danger to you, you have a much higher chance of damaging the device if power is connected when things come apart, whether the unit is turned on or not. Pull the plug, yank the batteries.

- **Rule number two**   Remove everything you can before going for the screws. Battery covers, recording media (tapes, discs, memory cards), rubber covers over jacks, lanyards—if it comes off, take it off. There's no need to pull knobs from a front panel if you don't anticipate having to remove the panel, but that's about it. Everything else should go. And don't be too surprised if you later wind up having to take off those knobs after all.
- **Rule number three**   Never force anything. If the case won't come apart, or some corner seems stuck, there's a reason. Perhaps snaps are hiding on the inside of the plastic. Screws are sometimes hidden under labels and rubber feet. Run your fingernail over labels, looking for the indentation where a screw head might be lurking. If you find one, peel back the label just enough to get to the hole. Peel back the rubber foot near the stubborn area; just because the other feet aren't hiding screws doesn't mean this one isn't. You'll probably have to glue the foot back on later, but it's a necessary consequence of peeling the original adhesive.
- **Rule number four**   Don't let frustration drive you to make a destructive mistake. Even the calmest tech can get riled up when a recalcitrant patient tries his or her patience badly enough. The most common errors are to start moving too fast, to force something, or, in extreme cases, to smack the casing, hoping it'll loosen up. Bad idea! Okay, I admit it: I once threw a really nice, rather expensive pocket stereo cassette player against the wall after a maddening, futile hour of trying to take it apart, but I don't recommend the exercise. It felt good, but, needless to say, that repair job was over before it began. Plus, for weeks I was stepping on pieces of that poor little thing I'd murdered. I couldn't help but think of Poe's *The Tell-Tale Heart* every time something went crunch underfoot. At least it was my own player, and not something I'd have to replace for a customer or lose a job over.

Separating Snaps

Popping apart hidden snaps is almost an art form in itself. First, be absolutely sure you've removed all the necessary screws. Take a look at the bottom of the unit to see if there might be slots into which you can put a screwdriver to pop the snaps. Those are common on AC-operated devices whose bottoms face shelving, but not on pocket toys. If you do find slots, shine a flashlight into one and see if you can deduce what needs to be pressed in which direction to unhook the snap. Pop one open while pulling the case halves apart with your free hand. To prevent accidental reinsertion, keep holding them apart while you do the next snap on that side of the case. Once you have a couple of snaps open, you won't need to continue the forced separation, and the rest should open easily.

If you find no slots, pick one side of the case and press on the seam around the edge, looking for inward bending of the plastic. Move slowly and feel for slight movements indicating where hidden snaps may lie. When the plastic gives a little, press harder and attempt to pull the seam up. If it won't budge despite your best efforts, move to another part of the case and try again. After you get one snap undone, the rest will release a lot easier.

If no amount of effort will release the snaps, you might be tempted to slip a small screwdriver into the seam and pry it open. It's a last-ditch procedure, but it usually works. It's almost certain to break snaps, though, and cause some visible damage on the outside of the case.

Even with the best technique, you'll break a few while you get the hang of it, and occasionally even after you're an expert. Luckily, it's not the end of the world. Often the loss of a snap or two has little or no effect on a product's integrity, but sometimes the reassembled case can feel loose or have a gap along the desnappified seam. Just save the broken pieces in case you need to melt them back on later.

Removing Ribbons

Ribbon cables have replaced wires in small products. They offer much higher density with a lot less mess, and we couldn't have today's complex pocket devices without them. The ribbons are delicate, though, and removing them from their sockets requires care.

Some ribbon connectors have latches that press the ribbon's bare conductor fingers against the socket's pins. Others have no latches and rely on the thickness of the ribbon to make a firm connection. Either style may have a stiff reinforcement tab at the end of the ribbon, but the latchless style always does.

Before pulling out a ribbon, take a Sharpie marker and put a mark on both the ribbon and the socket so you'll know how to orient the cable during reassembly. Use a Sharpie; other markers may rub off. If other nearby ribbons are similar enough that you could possibly confuse which goes where, use a unique mark on each one.

Now examine the socket closely. If you see small tabs at either end, it has a latch. Even without tabs, it might have a flip-up latch. Pulling a ribbon from a latch-type connector without opening the latch first can easily tear the ribbon. You don't want that! Ribbon cables are custom made for each product, and you aren't going to find a replacement unless you can dig up a dead unit for parts. A torn ribbon usually means a ruined device.

If there is no latch, grasp the ribbon at the reinforced tab and pull steadily. Don't jerk or you'll almost certainly destroy it. Pull gently at first and then harder if the ribbon doesn't move. With some of the larger cables, you might be surprised at how much force it takes to get them out of their sockets. If you need to pull hard, hold the socket down on the board with your other hand to prevent ripping it from its solder joints. It's rare for a ribbon to need that kind of force, but I've seen it a few times. Whatever you do, don't pull on the unreinforced part of the ribbon; it won't withstand the stress.

If there's a latch, open it first. There are two basic styles: slide and flip-up. Slide latches have little tabs at the ends of the socket. A fingernail or the end of a flat-blade screwdriver will pull them open. It's best to open them at the same time, to keep the sliding part from getting crooked or breaking off, but gently opening them one at a time usually works fine.

Flip-up latches open easily with a fingernail. You can use a screwdriver, but be careful not to scrape and damage the ribbon cable. Especially with a very small connector, open the flip-up slowly and carefully, as they break quite easily. It helps to pull them from near their ends, lifting both ends at the same time, rather than from the middle. If you break one, keeping enough pressure on the ribbon to make a proper connection with the socket is next to impossible.

Pulling Wire Connectors

Larger items may have a mixture of ribbon cables and good old wire assemblies. Circuit board-mounted connectors for wiring rarely have latches. If they do, the latch will be large and obvious, something you squeeze with your thumb while pulling on the connector. Most wire connectors simply pull straight up. Reorientation is not an issue with these, but marking is still advisable if other nearby connectors could cause confusion. We've all been taught through the years never to pull on the wires when removing plugs, but that's what you have to do with these because the plug fits entirely into the socket, leaving nothing else to grab. Grasp the wires, pull steadily without jerking, and the connector should pop out.

Before you do, though, be certain there actually *is* a connector! Groups of wires sometimes terminate in what look like connectors, but the wires go right through the plastic and are soldered directly to the board. Obviously, you don't want to pull on those.

Layers and Photos

Remember those little cups I suggested you collect, way back in Chapter 2? Here's where you will use them. Unlike the simpler products of yesteryear, modern gadgetry is often built in layers. Perhaps the topmost layer is a display. Under that lies a metal shield. Beneath that is a circuit board, and there's another board under that one as well. Behind all of it is the battery compartment and a little board for connectors. Ribbon connectors join the layers, with several on each side of the main boards, and you can't get to the lower layers without removing the upper ones first. Sounds like a huge product, doesn't it? Perhaps a home theater receiver or a laptop computer? Hah! I just described a typical digital camera! Just wait till you try to get to the back of the mechanism in a pocket-sized MiniDV camcorder.

To reach the innermost spaces, the layers have to be stripped away in precisely the reverse order of their original assembly. And, naturally, the problem you're chasing is at the bottom layer, right?

Each layer is held by screws, and they're probably different sizes from those holding the next layer down. Some may even be different from others in the same layer. Not all screws always have to be removed; some only grip small internal parts, and unscrewing them may drop a nut or a washer deep into the works. Especially in

FIGURE 9-2 Arrow indicating screw removal

devices with motors or speakers, both of which use strong magnets, a lost metal part can get pulled in and cause real trouble later. Some companies stamp a little arrow by the screws that must come out, especially on the outside of the product's case, but that's not an industry standard. See Figure 9-2. If you do see the arrows, remove only the screws that have them. If there are no arrows, start with the screws near corners, and see if that frees up the case. No? Then you'll have to remove them all and hope for the best.

As you unscrew a screw, observe how it comes out. It should rise, indicating that it's screwed into a fixed object, not a nut. If it seems very loose as soon as you start to turn it counterclockwise, screw it back in again and see if it tightens down without slippage. If not, then there probably *is* a nut on the other side, and you don't want to unscrew it! Nuts are almost never used on the screws holding case halves together. How would the manufacturer tighten such a screw without having access to the nut? Once in a great while, you do find a nut or a little metal bracket on the inside, glued into a plastic shelf so it'll stay put after the case is assembled. It's rare, though, and found mostly on older gear.

When you remove screws, take a good look at each one after it's out. Pay careful attention to the length, comparing it to the last one. Very often, otherwise identical-looking screws are of different lengths, and putting one that's too long in the wrong hole when you reassemble the device can make it poke into something, causing a short or other serious damage.

As you take out screws from a layer, put them in one of those cups. If different lengths are used, make a quick drawing of which went where on a little piece of scrap paper or a sticky note, and put that in the cup too. Now put another cup into the first one, covering the contents of the lower cup. When you start on the next layer, put its screws into the open cup, and so on. If the device is especially complicated or has many layers, take a photo of each layer, with the stack of cups visible in the picture. That way, you'll know which set of screws goes with which layer—just count the number of cups in the photo. You might be surprised at how easy it is to lose track of that after the unit has sat in pieces on your bench for days or weeks. When the disassembly operation is complete, you'll have a stack of cups, with the screws from the last layer in the top cup. Place an empty cup in the top one to keep the screws securely inside, and put the stack somewhere safe and out of your way.

Opening a Shut Case

Let's take a look at some case opening procedures for common products, starting with the easiest and working up to the really challenging adventures.

Receivers and Amplifiers

Most shelf-style audio gear opens up with no hassle. You'll find four screws, two on each side, two or three smaller ones at the top edge of the back panel, and perhaps one to three on top, just rear of the front panel. Unscrew them all, put them in a cup, and the top should slide off. Often, you'll have to spread the sides slightly while lifting the back edge, as the front edge is under the top of the front panel.

VCRs, CD and DVD Players

Most VCRs open the same way, but there are some variations. Some have screws on the bottom instead of the sides. You won't find one on top behind the front panel, but you may find some on the back's upper edge. On many VCRs, you have to slide the top straight back before trying to lift the rear edge.

CD and DVD players require much the same thing, but the front edge may have a lip fitting into a groove on the front panel.

TVs and LCD Monitors

Today's flat-panel TVs and monitors usually unscrew from the back. The panels are recessed from the bezel around them, so you should be able to lay the set face down gently on your bench, after sweeping the table's surface and checking for anything that could stick up and put pressure on the LCD or plasma screen. You may find lots of screws of various lengths. Be very careful to note which go where, because under those screws is the back of the display panel! You *really* don't want to mix up the lengths and screw anything into that when you put it back together.

Older CRT-type TVs have screws at the top and bottom of the back. They may be recessed, requiring a long screwdriver. It's best to lay small sets face down on a pillow before removing the screws. Some larger units let you remove the back without upsetting the stability of the TV, but many require that you tilt them forward a little bit to get the back off. That can be disastrous if the set flops onto its face. If you can put the TV on the floor and tilt it against a wall only a few degrees, that's your safest option.

Just remember that the neck of the picture tube, which will be in your face when you pull off the back, is dangerous, both electrically and mechanically. Be extra-careful not to hit the tube's neck, and don't touch the circuit board at the back. Nasty voltages live there, and they can persist even after the set has been turned off and unplugged.

Turntables

Turntables are an old technology, but they enjoy a following among audiophiles, so there are plenty of them still around. Also, turntables are uniquely shaped and somewhat delicate, making them awkward to service.

A turntable's platter may be driven by a rubber wheel, a belt or a direct-drive motor turning at the same rate as the record. Most better turntables are belt- or direct-driven. To change the belt, first lock down the arm so it won't flop around and damage the stylus. Then lift the rubber mat on the platter and you'll see the motor spindle somewhere on the left side. Putting the new belt on requires lifting the platter straight up and out. Most come off without removing anything, but some have a retaining clip around the spindle.

For any other repairs, you'll need to get to the underside of the turntable, which involves laying it on its face. To do that, put it on a pillow arranged such that the weight of the unit won't be on the arm. Never put pressure on the arm assembly; the arm probably won't survive.

Before you flip the unit over, it's wise to take off the stylus and put it aside, as it's the most fragile, easily damaged element. Whack anything into it and it'll get trashed. The quickest and safest way to get the stylus out of harm's way is to pull the entire cartridge. Many later turntables use a "p-mount" cartridge that unplugs easily, with no individual wires and connectors to deal with. Some with the old-style mount have removable head shells. If you see a sleeve where the head meets the arm, it's probably a removable shell. Unscrew the sleeve and it should pop right off.

Once the stylus and/or cartridge have been removed, secure the arm with its retaining clip. Take a look at the back of the arm. If you see a little anti-skate weight hanging down on a wire, make a note of its setting and then remove it so it won't get damaged when you flip the turntable over. If the primary counterweight slides on and off, slide it off after noting its setting as well.

Remove the mat. Unless the platter is held on with a retaining clip, remove the platter too, so it doesn't fall off.

Many turntables are mounted on springs, so you need to hold the corners of the chassis as you turn it over or the machine can fall out of its base. Hold those corners, turn over the unit slowly, and place the turntable face down on the pillow, making sure none of the weight is on the arm. If there's a bottom plate, remove the screws securing it, and it should come off.

Video Projectors

Be sure the lamp is completely cooled down, and take it out *first*! The bulb represents most of the cost of the projector. Plus, it's fragile and contains mercury. Put the lamp assembly aside, far enough from the work that you won't drop a tool on it or knock it off the bench.

Most video projectors have screws on the bottom. After their removal, the top half will lift off. Some projectors are entirely snapped together, with no screws. Even with screws, there may be hidden plastic snaps.

On some units, the lens has plastic rings for focus and zoom that must be pulled off before the case can be separated. Pry the rings off with your fingers, avoiding the use of tools. If you must use a screwdriver, do so especially carefully.

There may be ribbon cables between halves to connect the control buttons and indicator lights. Separate the halves slowly to avoid tearing them.

As you remove the case from a DLP projector, keep your fingers away from the front of the unit, because the color wheel is just inside, and it's fragile. Putting any pressure on it is likely to result in its destruction. LCD units have no color wheels to worry about.

Portable DVD Players with LCD Screens

These usually have screws of varying lengths in the back. After you remove those, the back should come off, but make sure to have the unit lying on its face, because the laser sled assembly can fall out and tear its ribbon cables if you hold it in any other orientation. The assembly sits on rubber bumpers, and the back holds it in place. It's supposed to be loose, for vibration damping and skip resistance. See Figure 9-3.

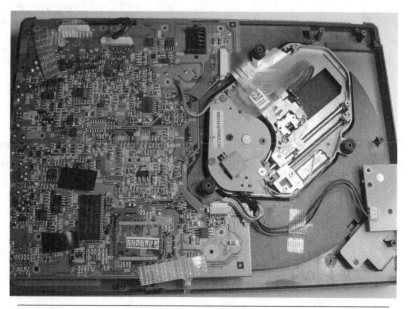

FIGURE 9-3   Inside a portable DVD player

If you need to get to the LCD monitor and its associated circuitry, including the backlight inverter, its screws are probably under the rubber bumpers on each side at the top. Check the back of the LCD before peeling off the bumpers, though, in case the screws are back there. On many players, the plastic bezel will come off the front, with the LCD anchored to the back, but some are the other way around. Often, the speakers are on the bezel, connected by wires, so remove it carefully to avoid tearing them. If the bezel won't come off after you pull the screws, it either has internal snaps or is glued at the seam. Many of them are glued to prevent rattling from speaker vibrations. Use the snap-popping procedure described at the start of this chapter. If you find no snaps, try gently peeling up the bezel one edge at a time. The feel of separating glue will be unmistakable. Just remember not to let the bezel pop off hard or you'll probably rip out those speaker wires. See Figure 9-4.

MP3 Players

These vary quite a bit, depending on who makes them. Some come apart easily, with accessible screws, while others are snapped together tightly and require a shim tool to separate. If you don't see a way in, get on the Internet and look for disassembly instructions. You'll find them, at least for the most popular players.

Flash memory players are usually just one circuit board, with the display mounted to it. You'll probably need to get to the troublesome headphone or power jacks, so you'll

Inverter

FIGURE 9-4 LCD with bezel off

have to remove the board unless all that's required is a resoldering of the jacks' contacts and the board happens to be oriented such that you can do the job *in situ*. If the player uses an internal battery, look for its connector and pull it before attempting repairs. As with other products, having power applied when you're working on the board can lead to circuit damage.

Hard drive players are a bit more complex, with a ribbon cable or two, and possibly more than one board. In most cases, it's best to remove the hard drive and put it aside. These units are more likely to have separate display boards, too, so be careful not to tear any ribbon cables when separating the case halves.

PDAs

These little handheld computers contain a remarkable number of parts, often on more than one board. You'll find a display, which also has the touch-sensitive digitizer, a rechargeable lithium battery, a main circuit board crammed with chips, and a switching-type power control and regulation system. Some PDAs have Wi-Fi and cameras, too, making for even more cables and boards.

The procedure for taking a PDA apart varies by maker, of course, but be prepared for a challenge. The keyword here is "small," so wear your magnifier. The multi-cup layer approach will be useful, as will taking photos as you go. Always keep the screen in mind, being careful not to scratch or press hard on it during your work. When you lay the product on its face, check your bench first for screws or parts that could damage the display.

Cell Phones

These are a lot like PDAs: lots of stuff in a very small space, with ribbons joining tiny keyboards and displays. Plus, there's an antenna on some that must be unscrewed and removed before the case can be separated. Be sure to take off the battery and pull the SIM (subscriber identity module) card as you begin. Be aware that the speaker magnet can attract tiny screws.

Camcorders

Camcorders that record to memory cards are like digital cameras, so see that section (next) for advice on taking them apart. Tape-based camcorders, analog or digital, are a whole 'nother story. These can rival laptop computers for complexity. In some ways they're worse because they are so oddly shaped that boards and mechanisms are crammed into nooks and crannies, making them hard to extricate. See Figure 9-5. Plus, the mechanism is delicate and easily damaged during disassembly. Drag out the cups and the digital camera.

If you only need to clean the heads, open the cassette door and then remove its two screws at the top. They may be under rubber covers. Take off the door and you should be all set.

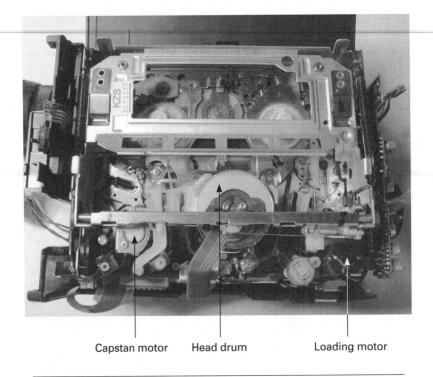

Capstan motor   Head drum   Loading motor

FIGURE 9-5   Mechanism side of a camcorder

For more extensive repairs, the machine will have to come apart. The typical camcorder body is in two pieces. Before you try to separate them, it's best to remove the cassette door, as it often prevents getting that side of the body off. If the machine works enough to get the door open, pop it open, take out its screws and remove the door. If the door is stuck closed, remove the screws anyway and see if you can slide off the door. Most likely, it won't budge, but you might be able to remove it once the case is loose.

Look for arrows on the case indicating which screws need to come out. You may find screws all over the case, and most of them will have to go. On some cameras, various covers on the front and top have to come off because there are screws under them securing the shell to the chassis. Gently pull the two halves of the case apart, being careful not to get your fingers inside, where they could damage the mechanism. You'll see ribbon cables all over the place, and you'll have to disconnect a few once the machine is apart.

Digital Cameras

These are some of the hardest items to service. Most of today's cameras are very slim and small, and the works are crammed in there tightly. Plus, cameras have lots of buttons, and some have sliding switches with plastic parts that fall off into oblivion as the case comes apart.

FIGURE 9-6   Inside a digital camera

The case halves on many digital cameras are three-dimensional puzzles. To get them apart, you may have to bend them around the edges slightly. On some, there's a plastic side piece surrounding the two halves, with tabs from each half fitting into it.

Generally, the back comes off, with all of the circuitry and the LCD remaining on the front. Be careful not to press on the LCD once the protective plastic window lifts off with the back of the case. See Figure 9-6.

As mentioned awhile back, digital cameras store the energy for the flash tube in a large electrolytic capacitor. That baby can hold its charge of several hundred volts or more for weeks after its last use. Usually, the cap is stuffed under the main circuit board, next to the optical assembly. See Figure 9-7. Its leads, however, may join the board just about anywhere. As you get the case apart, keep in mind that the connection to the flash cap could be right under your fingers. I've gotten zapped by digital cameras more often than by anything else. In addition to the danger to you, discharging the cap through your finger or a tool can leak high voltage into the camera's sensitive circuits, causing instant, silent damage.

Laptop Computers

Laptops are among the most complex consumer devices and the toughest to take apart. Talk about layers! You'll want to use the cups and camera for these. Before you begin, do an Internet search for disassembly instructions. Laptops are pretty trouble-prone,

FIGURE 9-7 The evil flash capacitor: note the voltage rating!

and sites abound with help. Very often, the disassembly sequence must be followed exactly or the machine can be damaged. Plus, where you begin depends on what area you need to reach. Changing the hard drive might require a different procedure and degree of disassembly than would resoldering the power jack or swapping out the backlight inverter.

Take off the battery before doing anything else. To remove the keyboard, look for snaps at the top. If you don't see any, check for screws on the bottom of the machine. They'll nearly always be placed such that they screw into the back of the keyboard near its top. I've seen a few near the middle, but none at the bottom, which usually has slots fitting into grooves on the top half of the case. Once the keyboard is loose, pull it up gently, keeping in mind that a ribbon cable connects it.

If you're trying to repair a backlight problem, check on the Internet to determine where the inverter is before taking anything apart. Sometimes it's in the body of the machine, but it's more likely to be in the LCD housing. If it is, you might not need to open the rest of the laptop at all.

Most LCD housings are screwed together. Look for screws along the edges of the housing. If you don't find any, check for cosmetic covers or bumpers on the front, near the bottom of the LCD. I've seen a few cases where screws were under the rubber bumpers at the upper corners, but not many. Often, those bumpers are pretty permanently attached and will tear if you try to pull them out. Then you find there's nothing under them anyway. Before going that route, exhaust all other possibilities.

Hidden snaps are common here too. Just avoid pressing on the screen; it's easy to do while pushing the seam along the edges, feeling for snaps.

Probably the most common failure in laptops is a loose, intermittent power jack. Repair is simple: just resolder the jack to the motherboard. Alas, getting to it isn't always so easy. For this one, you'll need to take the case apart. Look for screws all over the back, and keep track of their lengths when you take them out. Watch for hidden snaps along the sides, and don't bend the back too hard or you can break the internal frame or the motherboard. Go easy.

Some models mount the board to the internal frame, while others have it screwed onto the back. Netbooks and lightweight notebooks don't always have frames. Instead, the major components are simply screwed to the back.

If you're lucky, the jack's solder connections will be visible, and you can resolder them without further disassembling the machine. If not, you may have to remove the top half of the case. Be very careful of ribbon connectors going to the trackpad and other items on the top half.

I recommend not trying to take apart a valuable laptop if you've never done it before, as the chance of wrecking it is substantial. Outdated machines are available for very little, or even for free, from online resources and local computer recyclers. Get one and practice on it.

Chapter 10

What the Heck Is That?
Recognizing Major Features

When you open a modern electronic product, the apparent complexity may seem overwhelming. Even pocket-sized gadgets can sport a surprising complement of goodies inside. MP3 players and GPS units, for instance, are miniature computers, with RAM, ROM and a microprocessor. Some even have hard drives.

Older products, including those with double-sided circuit boards, usually had components only on one side. Not anymore! Thanks to the complexity and size of today's gadgetry, parts are mounted on both sides. Components that stick up, like transformers and can-style electrolytic capacitors, are often relegated to one side so the board can fit flush against the case. Everything else is fair game. Transistors, chips, resistors and small inductors and capacitors may be anywhere. As you wend your way through a circuit's path, you can expect to flip the board over numerous times.

To find your way around in a box crammed full of parts, wires and boards, you need to become acquainted with what the major sections of the product look like, how components tend to be laid out, and how to follow connections from recognizable features back to those less obvious.

Though the features vary a great deal depending on the product's function, pretty much every device has a power supply section, an input section (or several), some kind of signal processing and one or more output sections. Let's look at some common circuit sections and how to locate them.

Power to the Circuit: Power Supplies

Everything has a power supply of some kind. It could be a pair of AA cells, a simple linear supply or a complex switching supply with multiple voltage outputs. Power supply problems account for many repairs, so recognizing the power supply section is vital.

Batteries, obviously, are hard to miss. Battery-powered devices may have other power supply components as well, though, such as a switching converter to step up a single AA cell's 1.5 volts to a level high enough to run the product. Even when the battery voltage would be adequate on its own, a device may include conversion and regulation to ensure that a weak battery doesn't affect the reliability of writing to memory cards or hard drives, both of which can be seriously scrambled by insufficient voltage during write operations. Digital cameras and laptops have such systems so they can operate properly until the battery is nearly dead and then shut down the device gracefully.

To find voltage converters and regulation systems in battery-powered gear, look for small inductors and transformers. There are many varieties of them, but the telltale sign is metal. These things don't look anything like transistors and chips. They may be round or square, but most are made from ferrite material, which is a darkish metal with a matte surface. Toroid (doughnut-shaped) cores are common. You may be able to see the wire wrapped around the core, but don't count on it. See Figure 10-1.

Along with these, keep an eye out for electrolytic capacitors. Although 'lytics can be sprinkled throughout any circuit, the larger ones tend to be congregated in or near power supply sections. Diodes and voltage regulators will be found there too. Most very small products don't use enough current to require heatsinks on regulators and switching transistors, but larger devices usually do.

It might seem natural that you could follow the wires from the battery compartment straight to the supply section, but it doesn't always work out that way. Much of today's gear uses transistor switching driven by a microprocessor to turn itself on and off, rather than a real power switch actually interrupting the current between the batteries and the circuit. There may be some distance from where the batteries connect to the board and the location of the voltage conversion and regulation circuitry. Those little transformers are your best landmarks.

FIGURE 10-1   Miniature inductors and transformers

In AC-powered products, finding the power supply is a lot easier. Follow the AC cord and it'll get you there! Hard switches are still used in some AC devices, but those with remote controls, like VCRs, DVRs, TVs and some projectors, use the same microprocessor-controlled soft switching found in battery-op gadgets, so you can't count on the on/off switch's being in line with the AC cord. Look for a transformer. Switching supplies are pretty much standard now, but some products, especially high-end audio amps and receivers, still use linear supplies for their essentially noiseless operation. The transformer in a linear supply is a lot larger than the one in a switcher. See Figure 10-2. In a switcher, look for a transformer like that shown in Figure 10-3.

The power supply sections of pocket-sized devices are likely to be on the main circuit board, while those in AC-powered devices are almost always located on a separate board, with a cable feeding the output to the rest of the circuitry. See Figure 10-4.

Once you've located the power supply, its major features are easy to spot. In a switcher, the AC line will go through a line filter that looks like a small transformer, then through a fuse, and on to the rectifiers, which may be separate diodes, a bridge rectifier or a double diode, depending on the design. After that comes the chopper transistor, which is probably heatsinked. Near it will be a large electrolytic capacitor with a voltage rating in the hundreds. Then comes the transformer, followed by the low-voltage rectification and regulation circuitry. At the very end, right by the output wires, you'll see the large electrolytic filter caps with significantly lower voltage ratings than on the one next to the chopper.

FIGURE 10-2   Linear power supply transformer

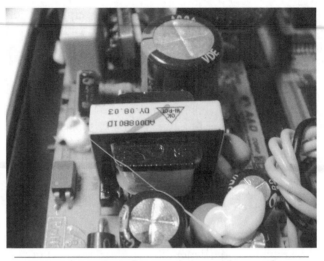

FIGURE 10-3 Switching power supply transformer

FIGURE 10-4 Power supply board. The chopper is hidden by the transformer.

Transformer 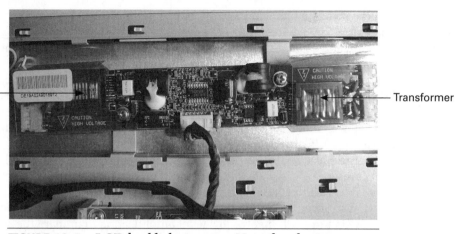 Transformer

FIGURE 10-5   LCD backlight inverter. Note the plastic insulation over the transformers.

In a linear supply, which has much less complexity, the AC line will go through the fuse and to the transformer. You won't see a heatsinked transistor or a big electrolytic cap on the AC line side of the transformer. On the low-voltage side, you'll find the rectifiers, regulators and filters.

A special type of power supply is the voltage inverter, a step-up supply driven from the main power supply's DC output. The inverter takes that low voltage and produces the high voltages required by LCD backlights. Inverters look like miniature switching supplies, which is basically what they are. The parts are much smaller, though, and you'll see two transformers and two output cables in bigger displays with two fluorescent lamps behind the screen. Most designs put the transformers at opposite ends of the board. See Figure 10-5.

Follow the Copper-Lined Road: Input

The input sections collect signals and feed them to the signal processing areas. The type of input circuitry present depends on the nature of the incoming signals. In radio and TV gear, input comes from an antenna or cable in the form of radio-frequency (RF) signals with strengths ranging from millionths to thousandths of a volt. The function of the input section is to amplify those very weak signals so further stages can separate out the desired one and demodulate it. The most common approach involves inductors (coils) resonating at the desired frequency. Analog tuners use a mechanically variable capacitor in parallel with the coil to change the frequency. Digital setups also use coils, but the tuning is controlled by the digital circuitry and accomplished with a *varactor*, which is a voltage-variable capacitor.

TVs may also accept baseband video, which is the analog video signal without an RF carrier. Various flavors of baseband video include *composite*, in which the entire signal is carried on one wire; *S-video*, in which the chroma (color) information is carried on separate wires from the luminance (brightness) signal; and *component*, which provides separate connections for red, green and blue. The input circuitry for each style of video is somewhat different.

The easiest way to find video input circuitry is to follow the lines coming from the input jacks. In sets with multiple input types, which is most of them these days, the lines will go to some sort of switching circuitry first so the set can choose the desired signal. After that, the signal will be sent to the appropriate stages for amplification and preparation for further processing.

Input can also be from a *transducer*, such as a laser optical head, phono cartridge, phototransistor, tape head or microphone. The signals from transducers may be very weak, as with magnetic phono cartridges, video heads and hard drive heads, or somewhat stronger, as with laser optical heads and phototransistors. Most of the time, the input circuitry will involve low-level amplification to prepare the signal for the signal processing sections. Because of their sensitivity to weak signals, input stages for transducers are often hidden under metal shields. Sometimes RF input stages are also shielded.

Shake, Bake, Slice and Dice: Signal Processing

Most products process a signal of some sort, be it analog or digital. Some kind of information is taken in or retrieved and massaged into whatever it is you want to hear, see, record, play back, send or receive. Much of the circuitry in any device is dedicated to signal processing. This is the little stuff, with lots of resistors, capacitors, transistors and chips. Digital circuitry is mostly chips, with a few bypass capacitors, and other small components that set operating parameters. See Figure 10-6.

FIGURE 10-6   Digital signal processing section

FIGURE 10-7   Analog signal processing section

Analog signal processing circuits also use chips, but those tend to be small-scale, with around a dozen leads, not a hundred. Analog sections also use more transistors than do digital ones, and you may see variable capacitors, potentiometers and adjustable signal transformers, especially in radio and TV receivers. See Figure 10-7.

While many products are primarily digital, plenty of them combine analog and digital functions, with a digital control system operating the analog sections. Frequency-synthesized radios use analog stages to pick up, amplify and detect radio signals, but their tuning is entirely digitally derived. Disc and MP3 players process everything in the digital domain and then convert the results to analog for output.

Digital control sections are recognizable by the large microprocessor chip with lots of leads. Look for a crystal or resonator very close to the micro. If the product has a display, it'll probably be near the heart of the control section, as will a keypad or a series of control buttons. See Figure 10-8.

Even in all-digital devices, signal processing, output and power supply sections look quite different from the control circuits.

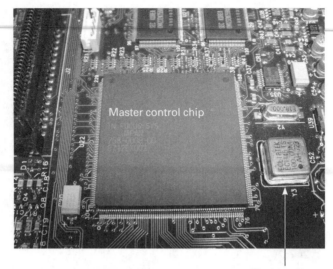

Clock oscillator

FIGURE 10-8 Digital control section

Out You Go: Output Stages

Output stages prepare the signal for display, a speaker, headphones, a transmitting antenna, a print head, a motor, and so on. In many cases, a large part of this preparation is current amplification, to give the signal the oomph to drive a speaker, move a motor or push an RF signal for miles. The primary identifying characteristic of output stages is that they are larger than most of what's around them. In some cases, like headphone amplifiers and small speaker drivers used in pocket radios and cell phones, the current required is low enough that the output stage may be quite small. Usually, however, more current is needed, so the parts are larger and better equipped to dissipate the higher heat.

The style of output stage depends on what is being driven. Audio amplifiers driving large speakers with many watts will have sizable heatsinks for the output transistors or modules. If modules are used, they will be much larger than power transistors, with more leads. If discrete transistors are employed, they'll range in size from around a postage stamp to perhaps 1 1/2 inches long. See Figure 10-9.

Output components used to drive motors may have heatsinks, but they might not if the motor is small and doesn't carry much load. Some motor drivers are just transistors on the circuit board. They're likely to be a little larger than the others, though.

FIGURE 10-9   These output transistors in a stereo receiver are larger than average, but they have only three leads, and their 2SC and 2SA part numbers prove they're transistors, not modules.

The output circuitry for dot-matrix LCDs is integrated into the display itself, along the edges. This stuff isn't serviceable; if a row or column goes bad, the LCD must be replaced. Simple numeric LCDs don't carry their own driver circuits. Instead, they're driven directly by the micro or by an external driver chip.

A Moving Tale: Mechanisms

We're heading toward a time when electronic products no longer include mechanical elements. Tape recording is already a dead technology. Eventually, everything will record to and play from memory chips, and even hard drives and optical discs will fade into obscurity.

That time has not come yet, and some of today's products still have mechanisms. Those in hard drives are sealed to prevent even the tiniest dust mote from crashing the heads into the platter, but the mechanics in optical disc players are readily accessible, and they cause enough trouble that you should get familiar with them. Digital Light Processor (DLP) TVs and video projectors use color wheel assemblies rotating at high speed. And if you're servicing older technology like VCRs, tape-based camcorders, audio tape recorders and turntables, you're going to get well acquainted with mechanisms and their peculiarities.

Figures 10-10 to 10-13 show some mechanical sections you'll encounter. For more details on specific mechanisms, see Chapter 14.

Capstan

Head tip   Pinch roller

FIGURE 10-10   Video head drum in a MiniDV camcorder

FIGURE 10-11   Camcorder capstan motor

Back of spindle motor   Back of laser head

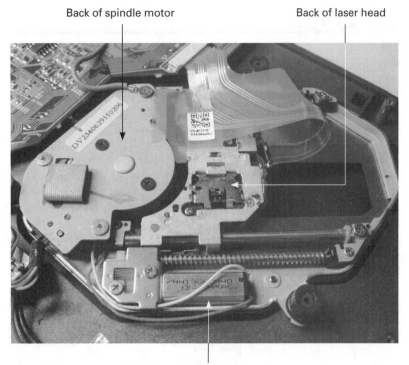

Sled motor

FIGURE 10-12   Laser optical head sled assembly

FIGURE 10-13   Color wheel assembly in a DLP
video projector

Danger Points

As we discussed in Chapter 3, there are dangerous spots in many electronic products. While it seems obvious that AC-powered circuitry would be the most hazardous, don't discount battery-operated gadgets as being harmless. Some products step up the battery voltage to levels that can give you a nasty jolt. Digital cameras, especially, generate hundreds of volts for their flash tubes, and it gets stored in a big capacitor capable of biting you weeks after being charged.

Watch out for any exposed points connected to the AC line. Also, heatsinks are usually grounded, especially in audio amplifier output stages, but those in switching power supplies may not be. The heatsink on a switcher's chopper transistor can be at hundreds of volts. *Never* touch it if the AC cord is plugged in. Even when the supply is unplugged, the heatsink may have a full charge from the electrolytic capacitor. Unplug the product and measure from the heatsink to the negative side of the big capacitor to see if any voltage is present. Remember, the AC side of a switcher is not connected to circuit ground, so measuring from the heatsink to circuit ground will show 0 volts, regardless of what's actually there!

The cases of metal power transistors in output sections can carry significant voltage as well. Avoid touching those without first measuring from them to circuit ground. Even 50 volts can do you harm if it's applied across your hands, especially if they are wet or sweaty. If you've ever touched your tongue to the terminals of a 9-volt battery, you know how little it takes.

The output connections of backlight inverters can be at 1000 volts or more. Keep away from them when the device is powered on.

VCRs, DVD players and some other products use small fluorescent displays. Lighting those up requires a few hundred volts, so beware of their connections.

Chapter 11

A-Hunting We Will Go: Signal Tracing and Diagnosis

Now that you have the unit open and ready for diagnosis, it's time to apply the ideas we've been examining and put that oscilloscope to good use. Locating the trouble is the heart of the matter and much of the battle. The general approach is to reduce your variables to eliminate as much circuitry as possible, concentrate on what seems a likely problem area, take some measurements, apply a little logic, and gradually narrow your focus until you reach the bad component.

Where to begin? That depends on what symptoms are being displayed. In order from least functionality to most, here are some good ways to pick a starting point.

Dead

As we discussed awhile back, dead means nothing at all happens when you try to turn the unit on. If that's the case, head straight for the power supply. Check the fuse first. If it's blown, assume something shorted and blew it. The short could be nearby, in the rectifiers, the chopper transistor or its support components, or it could be somewhere in the circuitry being powered by the supply, far from where you're looking.

If the supply feeds the circuitry through a cable, disconnect it. Replace the fuse and try applying power. Does the fuse blow again? If so, the problem is in the supply. If not, the fault could still be there, but more likely a short in the circuit being powered is drawing too much current and popping the fuse.

Some power supplies have small, low-current sub-supplies for standby operation, so they can keep enough circuitry alive to respond to remote-control commands or soft switches. The main supply turns on only when commanded to do so by the product's microprocessor. With the cable disconnected between the supply and the rest of the circuitry, the micro can't command the supply's main section to start up. If the short is in a part of the supply not running while in standby, the supply will appear to be okay and will not blow the fuse, confusing the matter of where the short lies.

If the fuse is not blown but the unit still does nothing, take a good look at the electrolytic capacitors in the supply and also in other areas of the unit. See any with even slight bulges on top? If so, forget about continuing your exploration until you've replaced them. By the time a cap bulges, it's pretty far gone, with perhaps 10 percent of its original capacitance left. Its equivalent series resistance (ESR) will be way up as well. Very likely, replacement of the bulging parts will restore operation of the unit.

If you don't see bad caps, check the supply's output voltages with your DMM. Find circuit ground on the output side (*never* on the AC input side of a switcher!) and hook the black lead to it. In a unipolar design, the negative output lead is almost always ground. If the supply is bipolar, there will be positive, negative and ground. Even if the supply provides several output voltages, one ground serves them all, although multiple wires may be connected to it.

In a metal-encased product like a disc player or VCR, the metal chassis should suffice as long as you can find a spot that's not painted over. In a pinch, you can usually use the outer rings of RCA jacks in audio/video gear. Choose an input jack, not an output jack, so you don't risk shorting an output if your alligator clip makes contact with the jack's inner conductor.

Many supplies have markings for the voltages on the boards, right next to where the output cable plugs in. If so, see if the voltages are there and are fairly close to their rated values. Don't worry if a line marked 5 volts reads 5.1. If it reads 4 or 6, then something's out of whack. When the voltage is too high, the problem will be in the supply's regulation. When it's too low, it could still be a regulator issue, but a short elsewhere in the circuitry might be pulling it down. If the voltages are okay, the supply is probably fine. If they read zero, it might still be fine and just isn't being turned on, as described, but it's quite possible it isn't working. If the supply is turned on and off by the unit's microprocessor, there still has to be some voltage from the standby supply to run the micro or it couldn't send a signal to start the main supply.

A product running off an external AC adapter might not blow its fuse even when seriously shorted. Most modern AC adapters are switching supplies. A well-designed one will go into self-protect mode, sensing the excessive current draw and shutting down. Usually, it'll restart every second or so, pumping some current into the device and then stopping again because the load is outside the normal range, never staying on long enough to melt the wire inside the fuse. Even the primary-side fuse in the adapter may survive, for the same reason. I once fixed a laptop computer's AC adapter that had a shorted output cable but never blew its fuse. The adapter's self-protect mode saved the fuse and the rest of the supply as well. The good fuse confused my diagnosis attempts until I considered the self-protect mode, checked the cable and found the short. After cutting off the bad section of cable and resoldering the remaining good length, I plugged the supply in and it worked fine.

Internal power supplies in AC-operated devices may also survive shorts without blowing their fuses, but they usually aren't as well-protected as external adapters, and the fuse blows.

If you have a working supply but no operation, head for the product's microprocessor and check for an oscillating clock crystal or resonator. If you find no voltage at all, there could be a little sub-regulator on the board to power the micro, and it might be bad.

If you see voltage there (typically 5 volts, but possibly less and very occasionally more) but no oscillation, the crystal may be dead. Without a clock to drive it, the micro will sit there like a rock. If you do see oscillation, check that its peak-to-peak (p-p) value is fairly close to the total power supply voltage running the micro. If it's a 5-volt micro and the oscillation is 1 volt p-p, the micro won't get clocked. If you have power and a running micro, you should see some life someplace.

Lots of products include small backup batteries on their boards. See Figure 11-1. These batteries keep the clock running and preserve user preferences. Loss of battery power causes resetting of data to the default states but doesn't prevent the product from working. In some cases, though, a bad battery can indeed stop the unit from turning on. I've seen laptop computers that wouldn't start up unless the bad backup battery was disconnected.

The batteries may be *primary* (nonrechargeable) lithium coin cells or *secondary* (rechargeable) types. Often they're soldered to the board. Primary types can be replaced with standard lithium coin cells of the same type number and a holder, as long as the arrangement will fit. You can even use bigger or smaller cells, since they're all 3 volts anyway; smaller cells just won't last as long. Secondary cells need to be replaced with the same type as the original, and those are not easy to find. Most likely, you'll have to try to order one from the manufacturer. Don't replace a rechargeable cell with a primary type, because the applied charging voltage will cause the nonrechargeable cell to burst.

If you're suspicious of a bad backup battery, measure its voltage. Should it be very low, disconnect the battery and see if the product comes to life.

FIGURE 11-1   Soldered rechargeable backup battery in a digital camera

Comatose or Crazy

This situation is trickier. When the unit turns on but is completely bonkers, with random segments on the LCD and improper or no response to control buttons, that usually indicates one of three things: power supply voltages are way off (probably too low), there's lots of noise on the power supply lines or the digital control system is seriously whacked out. Check the supply voltages first. If they look close to what they should be, scope the noise. Using your scope's AC coupling, look at the supply's DC output lines. There shouldn't be more than 100 mv (millivolts) or so of junk on them. If you see much more than that, you can expect to find bad electrolytic caps in the supply.

The most likely culprits are the caps right on the output lines. A good electrolytic will smooth out that noise, so its presence tells you the cap is not doing its job. Even if the part doesn't bulge or exhibit obvious leakage, try replacing it or temporarily putting another cap of the same value across it. Be sure to get the polarity right when you do this! And, of course, shut everything down first and verify that the existing cap is discharged. Don't worry about the two caps adding up to more than the correct value; some extra capacitance on a power supply line will only lead to better filtering. If the noise drops dramatically, change the cap, regardless of whether proper operation was restored when you jumped it with the good one. It might still be bad, but there could be others that will have to be changed before the unit will work again. If the noise drops only a small amount, then the original cap is okay, and you're just seeing the added capacitance smoothing things over a little bit. The trouble is elsewhere.

When you're sure the power supply is working properly, go to the microprocessor. Check it for clocking, just as with a dead unit. If the clock looks normal, it's possible that the reset circuit, which applies a pulse to the micro's reset pin when power is first applied, isn't working, so the micro isn't starting up from the beginning of its program. Many reset circuits are nothing more than an electrolytic cap between the positive rail and the reset pin, with a resistor going from there to ground. When the cap is discharged and the unit is turned on, the change in the cap's charge state momentarily lets a spike through to the reset pin until the cap charges up and blocks the rail's voltage. If that cap has dried out or leaked, the reset pin won't get tripped, and the micro will start up at some random place in its firmware, resulting in digital insanity. If you can find the reset pin, disconnect power, scope the line going to the pin and then reconnect power. You should see a pulse. If you don't, try turning on the unit. Still no pulse? The reset system isn't working. Try replacing that capacitor. Or, if there's a transistor, diode or other circuit generating the reset pulse, work backward, from its output to its input, scoping your way until you either find a pulse or locate its missing source.

Alive and Awake but Not Quite Kicking

This is where your sleuthing skills really get a workout. The unit powers up and responds properly, but some function doesn't work. Perhaps a backlight is out, or a disc player has trouble reading discs, or a VCR plays only in black and white. Maybe a projector turns on but the lamp won't strike. Figuring out these kinds of failures can take much more work than does troubleshooting the dead and semiconscious types.

After checking for the usual power supply issues, take an especially careful look on the board for bulging or leaky electrolytics. Change any that don't look normal. Sometimes 'lytics can be bad without showing physical signs. Scope them. As a rule, any electrolytic with one end tied to ground should have very little besides DC on the other end. Especially if you see high-frequency elements to the noise—it's fast or has spiky edges—jump that cap with another one and look at the noise content again.

If these simple checks don't turn up anything, it's time to go snooping. Is the problem at the input side, the signal processing midsection or the output stages? With audio gear, listen for a slight hiss from the speaker. If it's absent, the problem is likely to be in the output stages, because those will produce a little noise even if what's feeding them doesn't work. The speaker itself could be bad too; check with headphones or scope the output lines going to the speaker. With video equipment, there might be some noise on the screen, indicating that the stages driving it are working. With other kinds of devices, it could be harder to tell.

Items that move, like laser heads, print heads and swing-out LCD viewfinders on camcorders, have plenty of problems with broken conductors in their ribbon cables. If a moving part misbehaves, look at its ribbon with a magnifier. Even if you can't see a break, check the integrity of each line on the ribbon using your DMM. The very thin, flexible ribbons with black printed conductors rarely break, but the slightly thicker green ones with copper conductors are quite prone to fractured lines. Even if the conductor side looks black, check the fingers on the ends, where they make contact with the connector's pins. If they're copper-colored or look like they're coated with solder, check the ribbon carefully.

Sometimes Yes, Sometimes No

Want to give a tech nightmares? Sneak up behind the poor sap and whisper the word "intermittent." Watch for neck shivers and twitching muscles. Nothing is more difficult to find than a problem that comes and goes. Naturally, it goes when you look for it and comes back after you're done.

Thermal intermittents represent the easier-to-cure members of the genre. If a product works when cold but quits after warm-up, or the other way around, at least you can control those conditions while you hunt the trouble. Your most powerful weapon is a can of component cooler spray. After the operating status changes state, grab that can and get ready to spray. You don't want to spray the entire unit part by part, so concentrate on the kinds of components most likely to cause thermal problems: those that get warm. Voltage regulators, power transistors, graphics chips and CPUs all generate lots of heat and should show sudden change when you blast them with the spray, if they're the troublemakers. Spray small parts for about a second. Big ones may require as long as 5 seconds. Avoid hitting your skin, as the spray can cause frostbite. And, of course, keep your eyes away! I keep my face at least 12 inches from any part I'm spraying.

Now and then, thermal intermittents occur in small-signal parts too. If a transistor is leaky, it may get warmer than would a properly functioning part, driving itself into thermal weirdness. I remember one radio transceiver (transmitter/receiver) that would peg its signal strength meter to the far right when turned on, and no signals could be heard. As it warmed up, the meter would slowly drop to normal, and signals would gradually rise until they came in loud and clear. Knowing that the meters in radios are driven by the automatic gain control (AGC) circuit, I went hunting in that area. Spray can in hand, I finally found it: a small, garden-variety NPN transistor, leaky as could be. Oddly, it was shorting when cold and started working properly as the current through the short warmed it. Twenty-five cents and three solder joints later, the receiver was back to normal.

Electrolytics can be thermal as well. Of course, a cap shouldn't get hot in the first place. Some get a little warm, especially in switching supplies, but a really hot cap means there's a current path through it, so it's acting like a resistor. In other words, it's leaking.

If you blast any part and the unit starts or stops working, that's a pretty good sign you've found the bad component. When parts are crammed together, you may think you're hitting the right one, but a little bit of the spray is splattering on another component that's actually the culprit. Sometimes you have to spray a few times from different angles to be sure which part you're really affecting, letting the suspect and the components around it warm up again between sprays.

Solder joints can be thermal too. Sometimes when you spray a component and it starts (or stops) working, the real problem is at its joints. Before changing the part, always check the soldering and touch it up if you're not sure. Test again before replacing the component.

Mechanical intermittents are the hardest problems of all to find. When a machine exhibits symptoms by being tapped on, turned or tilted, there goes your night. And the next, and the next, probably.

Vibration or position-sensitive intermittents are caused by bad connections. They could be cold solder joints, circuit board cracks, dirty connectors, bad layer interconnects or, rarely, fractures inside components. Tap around, see what trips the symptoms, fix it, done. Seems simple enough, right? How hard could this be? Plenty. These kinds of intermittents tend to be very sensitive, causing malfunction no matter where you tap or flex. The basic search technique is to press and tap ever more gently as you home in on the problem area, hoping to localize the effect until you get down to one spot. Alas, even when you barely touch the board, the part flexing or vibrating may be far from your point of contact.

Circuit board cracks are rare these days, except when a product has been dropped. Most cracked boards stop working completely, but now and then a cracked trace will have its edges touching just enough to cause a vibration-sensitive intermittent. Far more common are bad layer interconnects, especially the conductive glue variety. Even plated holes can cause intermittents, but not very often. Conductive glue may look fine but not be making a solid connection with the upper or lower foil traces.

If you suspect a bad interconnect or a cracked trace, jumping with wire, even temporarily, will settle the question. If an interconnect isn't solid with an inner layer of the board, it can be tough to figure out where the jumper should go unless you have a schematic. Because the connection isn't totally lost, though, you can use your DMM to trace to other components. Keep an eye on the actual resistance to avoid reading through other parts and thinking they're connected when they're not. Expect to see some resistance. After all, that's the problem, right? If you see what looks like a connection, tap on the board and see if the reading changes. Remember that a DMM doesn't respond very fast. An analog VOM's needle will bounce, which is more useful in this case.

Many products use the chassis or case as circuit ground, with grounding pads on the board making contact when it's screwed down. As the device ages, loosened screws and oxidation degrade those critical connections, leading to intermittent behavior. In a unit more than 5 years old, check those pads even if the screws are tight. If the pads are dirty or oxidized, clean them up and see if that cures the symptoms. In a newer item, all should be well unless the screws are loose or the unit has lived in an especially corrosive environment like a boat.

Probably the most frustrating intermittent of them all is when the unit works just fine until you close up the case. Then it either won't work at all or it becomes motion-sensitive. You open the case back up again and the little monster works perfectly. Arggh!

To get to the root of one of these seemingly intractable dilemmas, consider what's happening when the case is closed. Look at the inside of the case and visualize where it will press on the board, on wiring and on ribbon cables. Some cases hold down the corners of the circuit board, flexing it when the screws are tightened. Experiment while the case is open, trying to re-create the problematic conditions. Most of these can be solved, but I've run into a couple I couldn't straighten out.

If the board isn't too sensitive to probe without altering the symptoms, use normal signal tracing techniques to locate the intermittent. If everything you touch disturbs the intermittent, it's very difficult to make sense of what you see on the scope.

To and Fro

Some techs like to work backward most of the time, starting at the output stages and hunting back toward the input area, looking for where the signals stop. Others prefer to start at the input and see where things get lost. What's the best method?

Either way may be appropriate. The output-to-input approach is especially useful when there is an output signal but it's not normal. Very often, such problems arise in the output stages and their drivers, so why start way back at the input and scope through stage after stage to get there? If you see a normal signal feeding the output stage, you've pretty much nailed it without a lot of hunting.

Digital devices offer a powerful clue to help you decide your direction of attack: is the time counter moving? If so, the device is receiving data, be it off a disc, a memory card or internal memory, and at least the heart of the digital section is working. So, start at the output and work your way back. If the counter is not progressing, head for the input area and find out why not.

With items like RF receivers, you may have normal audio hiss or video snow but no reception. Since the path through a receiver is fairly complex, with oscillators, tuned IF amplifiers and demodulation stages, it makes more sense to start at the input and work forward. At some point, you'll discover a missing oscillator or a dead IF or demodulator stage, with corresponding loss of signal.

If you are going to work backward, be certain you have a valid input before you start looking for it way down the line! Just because you plug an audio source into a stereo receiver, for example, doesn't mean it's getting to the amplifier board. There could be an issue with the input switching, or your connecting cable might be bad. Check the input signal *at the board* to be sure. Camcorder won't play in color? Are you sure the tape you're trying to play *has* color? If recorded on the same machine, it could be that the fault is in record, not play. Use a known good tape, or verify the existing one by playing it on another machine.

In many cases, a hybrid approach is the most effective. Start at the output and work back a few stages. If you can't find the signal, go to the input and work forward. In complex systems with multiple inputs, such as servos, check the inputs to be sure they're all there, since one missing signal will turn the whole thing into a mess.

All the World's a Stage

Always remember the all-important organizational concept of the circuit stage. You're not going to scope every darned component in the device. Instead, you'll focus on a particular area in the unit and look at it stage by stage.

Test points are very handy. With a schematic, you can look up TP204 and find out what it's supposed to show. Even without a diagram, you can often guess the signal being tapped from the waveform when you scope it. Sometimes you really get lucky, and the test points are labeled for function, in addition to their call numbers. You might see "reset" or "trk gain" (tracking gain). Checking those points and interpreting their signals can save a heck of a lot of work. If a test point at the end of a chain of stages shows the expected behavior, there's no need to scope each stage; they all have to be working.

Test points for digital signals like "reset" may show a line above the word. That means "*not* reset," which is tech-ese for "the signal goes low to initiate the reset, not high." When there's no line, the signal should go high, but don't count on that. Some manufacturers don't bother adding the line. If you see one, though, the signal definitely goes low.

Only when you find a nonfunctional stage is it worth trying to discern what part in the stage is preventing it from working. In the vast majority of cases, that part will be either an electrolytic capacitor or an active element: a transistor or a chip. With a few exceptions, like crystals and high-voltage transformers, other components that may have gone bad are probably victims of having had too much current pulled through them, and are not the perps themselves.

Diodes, rectifiers and zeners represent a special case. Though they're not active in the sense of having gain, they are semiconductors susceptible to the same kinds of failures found in transistors. Most techs think of them as active elements and check them before looking at more reliable components like resistors, coils and ceramic capacitors.

Zeners, which dissipate excess power as heat, are particularly prone to being open. Replacing a marked zener is no big deal because you can look the value up by its part number. Unmarked zeners present a much bigger problem if you don't have the schematic. What was the zener voltage supposed to be? You'll never know for sure, but you can make an educated guess.

First, the zener voltage will be less than what you're measuring at the blown zener, since the whole point of a zener is to reduce the voltage to the diode's breakdown rating; the part does nothing when the voltage is below that value. Theoretically, the zener voltage could be as little as a volt less than the applied voltage, but expect it to be at least a few volts less. Look for electrolytic caps in whatever circuit the zener regulates. The zener voltage will be less than the caps' voltage ratings. Again, it should be at least a couple of volts less, since few designers are foolish enough to run 'lytics all the way up at their ratings.

Though there's a wide range of zener values, many circuits operate on 5, 6, 9 or 12 volts, and it's reasonable to expect most of the zeners you find to be one of those values. Microprocessor circuits commonly use 5 volts. In audio power amplifiers, zeners are used to establish bias, and calculating the correct value isn't simple. Luckily, you should have another channel in which to measure the voltage across its good zener.

Check, Please

When you find a stage that isn't functioning, don't be too quick to indict it. First, be sure it's receiving the power and signals it needs to do its job. You really can't blame the poor transistor if it's not getting voltage, if its bias is way off, or some other stage isn't turning it on or providing proper input.

Unfortunately, cause and effect aren't always so clear. When a signal or a voltage appears to be missing, it could be that the stage is receiving it but a bad part is shorting it to ground. Or, a coupling component could be open, preventing the signal from getting to the active element. How to tell?

If there's a resistor between the source of the signal and the stage you're examining, check on the other side of it. The current limiting of the resistor isolates the far side from anything happening at the suspicious stage. The bigger the resistance value, the more isolation you can expect. A few ohms won't give you much isolation—signals will be about the same on both sides—but a kilohm or more sure will. You should be able to see something of the original signal on the other side, even if it's reduced in amplitude. If not, then the stage on that side isn't sending it, and you need to move your hunt to that part of the circuit.

A capacitor can provide isolation for AC signals, but how much depends not only on the size of the capacitor but also on the frequencies involved. The higher the frequency, the smaller a capacitor it takes to pass it, so the less isolation you get for a given capacitance value.

Once you're certain the correct conditions have been met, it is reasonable to conclude that you have a bad component, and it's time to start checking them. Unless you see a leaking cap, head for the active element first. Even if you find a burned resistor, you can bet the active element pulled too much current through it.

Sometimes parts can be tested while they're in the circuit, but usually the effects of other components will confuse the measurement, and you'll need to pull the suspect part before you check it. Two-legged parts need only one lead disconnected. If one lead goes to ground, remove the other one and leave the ground side connected. It's easier, since ground lands are typically the biggest and hardest to desolder. Also, you can leave one test lead connected to circuit ground and will have to connect only one lead to the component.

Wick the solder out of the hole and bend the part up on its other lead. For a three-leaded component like a transistor, pull two leads, and be sure one of them is the base or gate lead, so other parts can't influence the sensitive terminal with added capacitance or noise pickup via the rest of the circuitry.

With some parts, especially electrolytics, it might not be possible to bend the component on one lead if the leads are too short. If you can wick the hole thoroughly enough that the stub of the lead moves freely within it, testing is possible without pulling the component. With your test probe, push the stub away from the walls of the hole, watching the test results as you do so. The readings should make it apparent when contact with the rest of the circuit has been broken.

You'll be amazed at how many times you're absolutely *sure* you've found the problem, you pull the part, and it tests out fine. It can be frustrating, but that's just the nature of the repair experience. Eventually you'll nail it, and it feels really great when you pull the fifth part you were certain had to be the culprit, and it actually is!

When All Else Fails: Desperate Measures

No matter how good a sleuth you are, sometimes nothing works. You pull part after suspicious part and they're all good. You've been at it for hours, you're out of ideas and desperation sets in. Welcome to the technician's club! It happens to all of us once in awhile. Here are some desperation techniques to try. They may seem crazy, but they're better than giving up and tossing the unit on the junque pile. Now and then, they actually save the day.

Shotgunning

This is as old as electronics itself. When you have an intermittent connection you just can't find, solder them all! Back in the days when circuits had a few dozen parts, shotgunning was easy and quick. Today, with hundreds of joints on every board, shotgunning can take quite awhile, and it's not feasible with laptops and other extremely dense, complex products.

Start with an area you think is causing the trouble and hit every joint in it. If it doesn't work, keep on going. Don't be surprised if you wind up redoing every joint in the entire product, and it still won't work. Frankly, shotgunning is rarely successful; the real problem always seems to be something obscure that gets missed. Now and then, though, luck prevails and the symptoms disappear. Don't get too excited—you might have only wiggled the actual bad connection, and the problem will return... typically right after you tighten the final case screw. Once in a great while, I've seen shotgunning result in a real repair.

Current Blasting

This one has a little more basis in sanity. It's useful only when you have a dead short across the power supply rails somewhere on the board, but you can't find it. Especially on today's digital boards, there are lots of little bypass capacitors from Vcc (the positive supply rail) to ground. Now and then one of them shorts. You see the short no matter where on the rail you probe with your ohmmeter, so it's impossible to deduce which of the 50 little caps might have become a zero-ohm nightmare, and pulling them all to test them presents too much risk to the board. Plus, it'd take hours, and you can't be sure the short isn't in some other component anyway.

There exist exotic ohmmeters that read ultra-low resistance values down in the milliohms (thousandths of an ohm), allowing you to follow the traces and see when you're approaching the short. You don't have one of these babies, though, and neither do I. Even if you did, it would be hard to check the whole board and make sense of what you see.

There's a faster, easier way. You probably have a high-current power supply, either on your bench or perhaps in a discarded desktop computer. To perform current blasting, you need a supply of the same voltage as the product's supply. Many of these direct-short situations involve 5-volt digital boards, so a computer supply is a good choice. You need a *lot* of current—perhaps 20 amps or so. Your little 2-amp variable bench supply won't do it, but a PC supply has the required oomph.

If at all possible, disconnect the product's own power supply so you won't be feeding voltage into its output. That's usually okay, but some voltage regulators can be damaged when their output voltage exceeds their input voltage (which will be zero in this case), so it's best to avoid having them connected during this maneuver. If there's a removable fuse between the supply's output and the rest of the board, pulling that should do it. Otherwise, yank the connector or unsolder the positive wire.

With the hefty supply turned off, connect its +5V and ground wires to the board's supply rail and ground traces. Naturally, + goes to +. Make sure you're past any fuses, because this procedure will blow them. This is one time you don't want protection. Turn on the supply and wait. After perhaps 30 seconds, the shorted part will start smoking and burning, because pretty much all the supply current is going through it. As soon as you see the smoke, kill the supply; you've found the bad component. If the part is something nonessential, like a bypass cap, the product may start working as the current cooks the cap. I've seen the voltage rise high enough to start up a device even before the short clears. It's amazing what enough current will do.

This procedure will work with other voltages too, of course, as long as you have a supply that can source a lot of current.

Some caveats: it's possible the board's traces could melt before the shorted part gets hot enough to smoke. I haven't seen that happen, but it could. On a dense or multilayer board, a melted trace could prove disastrous. Also, the big supply must not have self-protection or it'll refuse to dump lots of current into a short. I've had good luck with desktop PC supplies; they are very sturdy and don't mind the overcurrent, at least not for the period of time required. They also don't seem to have self-protection circuitry.

If nothing gets damaged, current blasting pretty much always works when the shorted part is a capacitor. Sometimes the short is inside a chip, and the high current instantly blows it open. Nothing smokes, you don't know where the short was, the device still doesn't work, and you're no better off than when you started. Still, it's a useful technique and it beats just giving up.

I once fixed a really nice little hard-drive MP3 player that way. It had a dead short across the power supply input jack, and the board was too small and dense for me to consider trying to pull parts and test them. It was a 5-volt unit, so I hooked up a computer supply and hit the switch. In 10 seconds a surface-mount electrolytic cap right next to the power jack lit up like a tiny light bulb. I changed it and the unit came back to life. I saved a rather expensive digital piano with current blasting too. The shorted component was a tiny bypass cap near the microprocessor. I'd never have found it any other way.

LAP Method

This is the craziest last-ditch method of them all, but it has worked for me on rare occasion. LAP stands for "least accessible place." Where's the hardest place to reach in the entire product? If every other option has been exhausted, head there and suffer through whatever it takes to examine that difficult area. After a few LAP successes, I used to wonder how this could possibly be real. Was some cosmic force hiding things from me? Was there a ghost in the machine with a bad sense of humor?

The more reasonable explanation was that it was the one place I hadn't been yet! It seems like no matter how hard we try to check everything, there's always some forgotten nook so inaccessible that we don't even notice it, or we subconsciously avoid it. And, if it's the last possible place, the trouble just might be there.

Chapter 12

Presto Change-O: Circuit Boards and Replacing Components

Once you've found a component you want to test, or one that's obviously blown, you need to remove it from the board. Back when all components were mounted on leads pushed through holes in single- or double-sided circuit boards, removal was easy. A little solder wick or a pump of the solder sucker, and the holes would clear. After that, all you had to do was pull.

Sometimes the process is still like that, but now there's much more variety of component styles requiring different removal techniques, and multilayer boards have complicated the situation. Component removal ranges from trivial to maddening, and it's easy to destroy the circuit board when a recalcitrant part simply refuses to budge.

Unless both sides of the board are accessible, you'll have to remove it from the unit before you can desolder anything with leads poking through the board. Either way, first make sure power is disconnected. I always look at the AC plug before beginning to unscrew a board or desolder components, just so I know the plug is definitely lying loose. Even if I remember having pulled it, I take another look.

Through-Hole Parts

Many larger components still use the old wire-through-the-hole mounting technique. To remove power transistors and other through-hole parts, the solder must be sucked out of the hole, or the lead has to be pulled out while the solder is molten. Clearing the hole is preferable. For small joints, use solder wick, as described in Chapter 6. Place the end of the wick on the joint you want to desolder, and then press the iron's tip on the other side. Hold it there for about 20 seconds, and the solder should flow up into the wick. See Figure 12-1.

This doesn't always work, though. Sometimes the solder won't flow well enough to clear out the hole. The usual reason is insufficient heat, but transferring the heat to the joint is an issue too, as is thermal absorption by large copper lands. If you can't get

FIGURE 12-1 Using solder wick

a *small* land's hole to clear, try adding some fresh solder, and then wick it out again. Boards manufactured with lead-free solder don't desolder well. Adding leaded solder to a lead-free joint lowers the existing solder's melting temperature, making removal easier.

The wick absorbs some heat too, so it takes a hotter iron to desolder a lead than it does to solder it. Plus, to remove a lead requires wicking out all of the solder in the hole. With thick or multilayer boards, some of it may be a millimeter or more away from the heat source, making the solder hard to melt.

Desoldering is complicated by the increased thickness of multilayer boards and their extra heatsinking effect from internal foils contacting the copper coating inside the holes. Applying enough heat to wick the solder out can destroy the board. To remove a stubborn lead from a multilayer board, it's best to heat one side while pulling the lead out on the other, and then clear the hole after the lead is gone. Even then, you may struggle with it and be tempted to reach for the big soldering gun. That's too much heat for small boards, and it can deform them and break internal connections in multilayer boards, wrecking the device. See Figure 12-2.

Large lands used for power supply and ground buses create a heck of a heatsinking effect. It can be quite frustrating trying to get them hot enough to melt and clear the solder. The big gun might be called for here, but it's still possible to trash the board because there may be other lines running over the big land inside the layers. Heating up the big land can break them or short them to their adjacent layers.

If a part won't come out no matter how hard you try, it's a lot safer to clip the leads and solder in the new part without clearing the holes. Clip the new part's leads close and solder them to the residual solder in the holes. You should be able to heat a hole enough to make a good joint, even if the solder at the far end of the hole never melts. If there's room, you can leave a little of the old part's leads and solder to those.

FIGURE 12-2   How *not* to do it! Excessive heat destroyed this multilayer board.

Sometimes you can't get to the leads to clip them. On most electrolytic caps, the leads are under the parts, unreachable with any tool. The easiest way out is just to chop off the component near its base with a pair of wire cutters. Then you can clip or desolder the leads easily.

Bigger joints with lots of solder can overwhelm solder wick, saturating it before much solder is removed. To clean out an entire large joint might require a foot of wick, which isn't cheap. These are jobs for solder suckers. After applying the sucker a few times, you should be left with only a coating of solder on the joint. A sucker will not remove that, so finish up with wick.

As mentioned in Chapter 6, avoid using a spring-loaded solder sucker on static-sensitive components like CMOS chips and MOSFET transistors. The rapid release of the plunger can generate static charges capable of damaging those parts.

Surface-Mount Components

Wicking surface-mount parts is easier because all of the solder is touching the wick, and many of the lands are very small and readily heated. Most surface-mount pads will desolder without incident. If the solder on a small land won't flow into the wick, try the same trick I described above: add some fresh solder to the joints before trying to desolder them.

Large lands on power supply and ground buses may still be hard to heat, but a normal iron will take care of most of them. Using a big gun on a surface-mount component is asking to destroy the part and quite possibly the board. Tiny SMT (surface-mount technology) resistors and capacitors have sputtered-on solder pads. Too much heat can delaminate them, making reconnection to the parts' bodies impossible to achieve.

Most SMT components are glued in place before being machine-soldered at the factory. Very often, desoldering the ends of a part will break the glue and free the part, but not always. If you see a red shellac-like blob around the edges of the component, it's glued on and may not budge after desoldering. To move it, wick both ends and then heat one end while pushing on the component's body with a small screwdriver. The tiny part may pop off suddenly and blast away into oblivion if you're not careful. Somewhere in the universe there must be a room full of sad, homeless SMT parts that flew off circuit boards, never to be found. Plenty of 'em came from my workbench.

Choosing Components

Any time you need a new part, you just breeze on down to your local electronics supply store, buy the exact replacement and pop it in. Um, right, sure you do. Ah, if only real life could be like that! We don't even have local parts stores anymore. And while lots of standardized components are available via mail-order, many newer consumer electronics products aren't made from them. Instead, they're stuffed with all kinds of obscure and specialized components nobody but the manufacturer can provide. Luckily, in most cases you have a few options.

Ye Olde Junque Box

If you've stockpiled components, see if what you have is a close enough match. When using parts that have been sitting around for a long time, take some fine sandpaper to the leads to remove oxidation that will have built up. Otherwise, soldering to those leads will be unsuccessful.

Parts Machines

There's a reason I've encouraged you to save boards from dead machines. Those from the same manufacturer as the unit you're repairing might use the same component, even if they're a different model. Manufacturers save costs by reusing parts of their designs and techniques in lots of models. If you can't find an exact replacement, you still might locate something close enough to work. Check all your parts machines, even those made by other companies. You're more likely to find a compatible part from the same type of machine, since the function is similar. So, if you need a part for a camcorder, check boards from those; you probably won't find what you need in a DVD player. If you locate something you can use, but the leads are too short, solder on a little wire to extend them.

Substitutes

Substituting a part with something close but not an exact match requires consideration of how the part is being used, what parameters are critical, and what you can get away with in a particular application. The general idea is that a part with better specs can sub for a lesser one, but not the other way around. Even then, there are exceptions. Different component types have varying requirements. Let's look at some common ones.

Capacitors

Most of the capacitors you'll replace are electrolytics. Tantalum caps fail pretty often too, but they aren't used much anymore, so you may never run across one. The major factors in an electrolytic are its size, its capacitance, its voltage rating and its temperature rating. Also, switching power supplies and computer motherboards often require caps with especially low ESR, to smooth out the fast, sharp pulses those circuits produce. Replacing such parts with standard electrolytics will cause malfunction.

The most important consideration after size—it does, after all, have to fit on the board in the allotted space—is voltage rating. Electrolytics simply won't stand voltages higher than their ratings, at least not for long; they fail catastrophically by shorting. Their life is reduced even by running them at voltages under but close to their ratings, yet some manufacturers will use a 15-volt part at 13 volts, leading to frequent failures. In cases like that, a replacement with a higher voltage rating than the original part is not only okay, it's desirable.

The capacitance rating is not as critical as you might suppose. Most 'lytics have rather wide tolerances, in the range of -20 to +80 percent. If the cap is being used to couple signals from one stage to another, the capacitance value is more important than it is when the part is a bypass or filtering cap. You might find a few electrolytic coupling caps in audio and video gear. In audio amplifier stages that use caps of a few microfarads from an emitter to ground, it pays to keep the value close to the original, because a higher value might increase low-frequency response, upsetting the audio quality.

You will see tons of electrolytics in power supplies and for bypassing and filtering in all kinds of products, from simple analog devices to today's most complex digital gear. Those are the parts that usually need replacement. If your available replacement's value is no more than 50 percent higher, go ahead and use it. A little extra filtering never hurt anything, and +50 percent is likely within the stated tolerance of the original part anyway. To be sure the new part isn't at its maximum tolerance value of, say, 80 percent over the stated value, measure it with a capacitance meter. Despite their wide stated tolerances, most electrolytics I've measured have been within ±20 percent or so of their printed values.

Combining capacitors to get near the needed value is fine in most applications. I don't recommend it, though, for the big storage cap at the input of a switching power supply (near the chopper), or in other high-voltage circuits. Putting caps in parallel adds their values, and putting them in series drops the final value according to the following formula:

$$\frac{1}{\frac{1}{C_1} + \frac{1}{C_2} + \frac{1}{C_3} \cdots}$$

Capacitors combine exactly opposite to how resistors do. For more info, see the section "Resistors" a bit later in the chapter.

When putting polarized capacitors in series, be sure they connect + to –, so you wind up with one + and one – at the ends of the string. When you parallel them, connect all the + terminals to each other and all the – terminals to each other. In either case, be sure each capacitor's voltage rating is equal to the entire applied voltage. When in series, the individual caps won't really be subjected to the full voltage during normal operation, but a big voltage spike can occur when power is first applied, so it's a smart safety move to be certain every one of them can handle it.

Especially in power supply applications, the cap's temperature rating matters. Electrolytics that get charged and discharged very fast, as they do in a switcher, can become plenty warm from the power dissipation of their internal resistance. Standard 'lytics are rated to operate at 85° C, with higher-temperature caps rated as high as 150° C. Manufacturers hate paying for things they don't need, so respect the temperature ratings if you want the repair to last. For quick testing purposes while troubleshooting, you can disregard the ratings because the part won't be running long enough to fail from overheating.

Tantalum capacitors should always be replaced with the same type. They have lower impedance at high frequencies than do standard electrolytics, and are used only where that matters. Replacing a tantalum with a garden-variety electrolytic will result in performance degradation or circuit failure. The capacitance tolerance of tantalums is much tighter than that of standard electrolytics, so use a part with the same value. An increased voltage rating is fine, however.

Diodes

Diodes and rectifiers have four primary characteristics: forward voltage, reverse voltage, current and speed.

The forward voltage spec tells you how much voltage can be across the part in its conducting direction. You won't often see this specified, because in an AC circuit the forward and reverse voltages are usually the same. The reverse voltage is specified as *PIV*, for peak inverse voltage, and it tells you how much voltage the diode can withstand in its nonconducting direction. Exceed the PIV, and the part will arc over inside and be destroyed.

The current rating indicates how much current can pass in the conducting direction without overheating the part and burning it out. No current should flow in the reverse direction, of course.

The speed of a diode is very important in some small-signal applications like radio signal detection. It's also critical on the low-voltage side of a switching power supply, where the part will be rectifying the fast pulses from the conversion transformer. In the sections of power supplies operating at the low frequency of the AC line, any rectifier is more than fast enough. The bridge or individual rectifiers at the AC cord side of a switcher are not high-speed devices; nor are the rectifiers on the low-voltage side of a linear supply.

When subbing a normal, low-speed rectifier or bridge, pay attention to the PIV and the current rating. As long as those are equal to or higher than the original part's ratings, the new part should work fine.

Look up high-speed rectifiers in a substitution book or online. Replace them with parts of equal or better PIV, current and speed. Never replace a high-speed rectifier with a low-speed part, even if the PIV and current specs are fine. It simply won't work.

Some products use lots of glass small-signal diodes. Look for numbers like 1N914 and 1N4148. They're interchangeable. Even if you see no number on the diode, either of those numbers should work fine. Just be sure the diode you're replacing is in fact a simple diode and not a zener. There are some other special-purpose diodes, too, including germanium diodes (also glass but noticeably larger than normal silicon diodes), gallium-arsenide diodes, tunnel diodes, and varactors. They're found in receiver front ends and other weak-signal, exotic applications. You won't run into them very often, but they must be replaced with diodes of the same types.

Resistors

Many resistors are carbon composition types and easy to sub. What matter most are the resistance value and the power dissipation capability. It's fine to use a 1 percent precision resistor in place of a standard 5 percent one, and it doesn't hurt if the replacement is rated to handle more power.

If the original resistor was a special type, such as a wire-wound or low-noise part, it's important to replace it with the same type. Those kinds of parts are used only in special applications. You won't find them very often in consumer electronics gear, but they show up now and then in switching power supplies, preamps and stages handling particularly small signals, like receiver front ends.

If you can't find the exact value you need, consider the original part's tolerance (see Chapter 7), and try to combine a few other resistors to get to a value well within the original part's specs. For instance, if you need a 3.3 KΩ resistor, you could put a 2.2 KΩ and a 1 KΩ in series. Resistor values in series add together, so that'd get you to 3.2 KΩ. If the original resistor had a 5 percent tolerance, as most do, it could vary by ± 165 ohms and still be okay. So, 3.2 KΩ would be fine as long as the combined resistors' own tolerances didn't push their total value outside the tolerance range of the original part. Check the real value of the combination with your DMM to be sure.

Resistors in parallel combine opposite to how capacitors do. The resistance value goes *down* according to the formula shown here:

$$\frac{1}{\frac{1}{R_1} + \frac{1}{R_2} + \frac{1}{R_3} \cdots}$$

Two resistors of the same value will produce half the resistance. The larger the resistance of the second resistor, compared to the first, the less effect it has on it. Play around with a few resistors by combining them in parallel and measuring them, and you'll get the hang of it.

Transistors

Transistors are the most complicated parts to substitute. Major semiconductor manufacturers used to give away large transistor substitution books filled with hundreds of pages of transistor types and their brands' appropriate cross-referenced substitute part numbers. Because many types of transistors have similar characteristics, a few hundred parts can sub for thousands of parts.

These days, you can look up this stuff online, but you may run into numbers for which you can't find a cross, or there might be a valid sub but you can't get one. Alas, some parts are still made of unobtainium. Even when a substitute component is available, you may prefer to speed up the repair process by using a part you already have.

To choose your own substitute requires some understanding of the part's application and how a change in characteristics would affect circuit performance. Some functions, like simple switching of voltage to direct it to various circuit stages or turn an indicator on and off, will work with just about any transistor of the same basic construction (bipolar or FET) and polarity. Others, such as high-frequency signal processing or current amplification in complementary output stages, often require stringent adherence to the original part's specs.

All this assumes you know the old part's number. Usually you will, but at times you might have to fly blind. If the original transistor literally blew apart, which occasionally happens when a heck of a lot of current has been pulled through one, there may not be a number to read! I've seen SMT output transistors in LCD backlight inverters blow so hard that there was little left between the solder pads. Even when the number is visible, it could be a proprietary house number, with no cross-reference to a sub. And some transistors, especially tiny SMTs, show no numbers in the first place.

If you're lucky, the board will be marked with ECB or GDS, showing what terminal goes to which pad. ECB indicates emitter, collector and base, thus a bipolar transistor. GDS means gate, drain and source, the terminals of a FET. Those markings also give you strong clues to the part's polarity. If C goes to the positive side of things, it's an NPN. If E does, it's PNP. With a FET, if D is positive, it's an N-channel part. If S is positive, it's P-channel.

Without board markings or a part number, the transistor is a total mystery. Use your scope and understanding of basic transistor operation to deduce the part's polarity and layout of connections. Start by looking for the power supply voltage feeding it. If it's positive and fed through a resistor or a transformer, you've probably found the collector of an NPN transistor or the drain of an N-channel FET. Find the stage's input by looking for whatever signal operates the transistor. If it's a continuous signal, you should see it. If it's something that happens only when you press a switch or some other operation signals that area of the circuit, create those conditions and find the signal. When you find it, you've found the base or the gate. Whatever's left will be the emitter or the source.

Most bipolar transistors are NPN. If the connection to the positive supply line is direct, without a resistor, or there is a resistor but it's of very low value, the transistor could be PNP, and that connection would be its emitter. Find what looks like the base by scoping for signals. See if there's a resistor from the base to the transistor terminal closest to the supply. PNPs are used to turn on and pass current from the supply to some other circuit when the input signal goes low, toward ground. The resistor going up toward the supply keeps the base high and the transistor turned off until the input signal pulls it low. You'll find PNP circuits of that sort in power switching sections of battery-operated products.

Assume the part is an NPN bipolar transistor, and you'll be right most of the time. If your replacement turns out to be the wrong polarity, the circuit won't work, but it shouldn't do any damage.

All bets are off if the transistor is part of a complementary push-pull amplifier. They use NPNs and PNPs in more complicated, hard-to-deduce ways. And if the original part was a FET, the issues of enhancement and depletion mode, and JFET versus MOSFET, make the whole thing very tough to fathom. Getting the identification correct requires your understanding how those parts work and looking at the bias on the gate terminal to infer what the output should do as the input changes.

Don't try to sub chopper transistors in switching power supplies without knowing the correct part number and finding a legitimate sub from a cross-reference. Most choppers are power MOSFETs with specs that must be closely matched for reliable operation. Even if a sort-of-close sub works, it probably won't run for long before failing. Sometimes even a legit sub will die in a hurry, and the only part that will work is an exact replacement of the original part number. The same is true of horizontal output transistors in CRT TVs, another application involving fast pulses at fairly high voltages and currents.

In some cases, the original and replacement transistors are electrically compatible but their arrangement of leads, called *pin basing*, is different. Most small-signal American transistors are EBC, left to right, while Japanese parts are usually ECB. You can replace one layout with the other as long as you switch the two leads, being careful not to let them touch as they rise from the board toward the transistor. Small FETs are usually SGD. Power transistors are usually BCE, with C connected to the metal tab (if there is one), but check to make sure. Power FETs typically use GDS, with D connected to the tab.

Once you've figured out what should go where, whether from the original part or from scoping and deducing, you can proceed with trying out a new part. The primary characteristics to be concerned with are gain, high-frequency cutoff point and, with larger parts, power dissipation capability. Secondary characteristics, but still very important, are the maximum voltages permitted from base to emitter and from collector to emitter.

Very often you'll find a transistor that's pretty close but has a little more or less gain. Depending on the application, that might work. If the circuit is linear, producing output proportional to the input, the transistor isn't normally saturated (fully turned on), so a slight gain difference may not cause a problem. In switching circuits like backlight inverters, though, inadequate gain can result in lots of heat from the transistor's not-fully-turned-on resistance, burning out the part in a hurry. Too much gain in a linear circuit may cause distortion, increased output or spurious signals. Not enough usually just results in a bit less output.

The high-frequency cutoff point specifies at what frequency the transistor's gain will have decreased to one. In other words, it won't be amplifying at or above that frequency. In low-frequency applications such as audio, any transistor will be more than fast enough. At radio frequencies, the situation can be quite different, requiring a transistor whose cutoff frequency is approximately equal to the original part's spec. Too little might result in low or no output, while too much could result in unwanted harmonics or spurious signals riding on the desired one. When in doubt, go for too much, as long as the difference isn't excessive; at least the thing will try to work.

Power dissipation is very important. The new part should be able to dissipate at least as much power as the old one. A better dissipation spec is fine.

Maximum permissible inter-electrode voltages must be respected. Exceed them and the transistor might emit some of that magic smoke. Most transistors' collector-to-emitter specs are well beyond what a small-signal circuit produces. The circuit's base-to-emitter voltage, however, could exceed the capabilities of some replacement parts, so keep an eye on that. Large parts used in output stages can have pretty high voltages applied from collector to emitter, so don't take that spec for granted.

If all this seems overwhelming, stick to replacement part numbers from a cross-reference book or online source, and you'll be fine. Even with expertise, matching up transistors is very much a roll of the dice. See, I told you it could get complicated!

Zeners

The purpose of a zener diode is to break down nondestructively in the reverse direction and conduct when the part's reverse voltage spec, or *zener voltage*, is reached. The important specs are the zener voltage and the power dissipation. Unlike normal diodes, zeners' dissipation limits are specified in watts, not amps. Always replace a zener with one of the same zener voltage and at least as much dissipation capability. A higher dissipation spec is fine.

You can put zeners in series to add their voltages, but don't parallel them to increase dissipation capability; even zeners with the same zener voltage won't start conducting at exactly the same voltage, so one will always take more current than the other, resulting in its premature failure. When combining them in series, be sure that the wattage of each zener is at least as high as the original part's rating, and watch the polarity. Each zener should feed the next one cathode to anode, so you wind up with one anode and one cathode at the ends of the string.

Installing the New Parts

Once you've procured or substituted components, it's time to put them in! Proper installation is crucial for successful, long-term repair. Let's look at some issues specific to various kinds of parts.

Through-Hole

Replacing a through-hole component is pretty easy, requiring nothing more than pushing the leads through the holes, bending the ends a little so the part doesn't fall out, soldering the leads and then clipping off the excess.

If the part is attached to a heatsink, it's a little more complicated, but not much. For a free-floating heatsink bolted or clamped to the top of the component, install the heatsink before soldering the part to the board. When the part mounts on a fixed heatsink, put the leads through the board's holes without soldering them, and then screw or clip the component to the heatsink. Solder the leads only after the mounting procedure is complete.

If the original part used heatsink grease, you need to do the same with the new one. The grease used is a special silicone compound formulated for maximum heat transfer. You can get it from online parts houses, and computer supply shops that carry CPU upgrades and bare motherboards also carry it. Most heatsinks, including those with mica or thin plastic insulators, do require the grease. Those with rubber separators usually don't, though. Figure 12-3 shows typical insulator setups requiring thermal grease.

A thin smear of the special grease on one of the mating surfaces helps heat transfer across the less-than-perfect contact area, filling in tiny gaps and increasing effective surface area. Too much grease can separate and insulate the surfaces, *reducing* heat flow, so don't overdo it. Smear on the grease with a swab, and be careful not to put bending pressure on the insulator or it may break. Mica insulators are especially brittle, and even a single crack can lead to a short later on. To avoid bending it, place the insulator on your workbench before applying the grease.

The insulator's job is to isolate electrical contact between the component and the heatsink while facilitating heat transfer. If there's no insulator, either the part

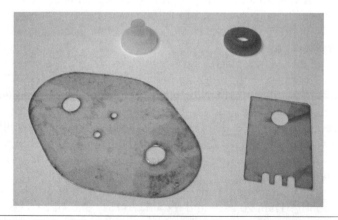

FIGURE 12-3   Transistor mounting hardware with screw sleeves

has no contact point on its case, as with an all-plastic transistor, or it's okay for it to be connected to the heatsink. Voltage regulators sometimes have their ground connections on the metal tab, so contact with a grounded heatsink is a good thing. Many power transistors, though, have their collectors or drains at the tab. Those are usually connected to voltage sources, and contact with ground would be a short. Insulators are used to avoid the connection.

When there is an insulator, and the component has a metal tab, the mounting screw will pass through a plastic washer with a sleeve. Be certain to use it, and watch its orientation. The sleeve should fit into the hole on the transistor's tab, preventing the screw from touching the inside of the hole.

Tighten the mounting screw more than you would a screw holding a board down or a case together. You want good heat transfer, and that takes some pressure. Don't overdo it to the point of breaking the insulator or stripping the screw, of course.

Occasionally, you will find a thermistor (a heat-sensitive resistor) glued to the case of a power transistor, especially in the output stage of a push-pull audio amplifier. Thermistors are used to adjust the bias of bipolar power transistors as the parts heat up, because their gain and optimum bias point drift with temperature. If you can get the thermistor off without destroying it, glue it with epoxy to the new part. If you can't remove it, you'll need a new thermistor. Look up its part number and order one just like it.

SMT

Putting in a new SMT part is a bit tougher than installing a through-hole component, thanks to the size scale. How do you hold it in place long enough for soldering? Gluing is not recommended. Sure, the manufacturers do it, but they have special glue made for the purpose, and we don't. More than likely, some other line runs underneath the component, and a later attempt to remove the glue will tear the copper off the board. Also, the electrical properties of the glue you might use are a wildcard; you have no idea how its presence might affect circuit performance. It could exhibit capacitance or even conduct current.

To get an SMT in place, first use wick to clean the board's solder pads so that there are no raised bumps of solder on them. You want the SMT to lie flat. Melt a little solder onto your iron's tip. Now place the part on the board and line it up carefully with a tiny screwdriver. Center it between the pads so it can't create a short across two lands. Hold down the body of the SMT with the screwdriver while touching the iron's tip to one end of the part. The solder on the tip should flow onto the board, making a joint at that end. Don't worry about getting a good joint; all you want to do is prevent the part from moving.

Once the component is held in place by the solder on one end, solder the other end properly. Then go back and redo the messy end. Take a good, close look with your magnifier to be sure you haven't created any solder bridges to adjacent pads or parts.

Finding Parts

Proprietary components have to be procured from the manufacturer (unlikely these days, but worth a try), the component maker who supplied them (possible) or a parts unit. For popular gadgets, finding a parts unit may be the easiest way to go. Check eBay.

Standard components are widely available through online mail-order, but many parts houses have minimums, so you might have to spend a lot more than the part is worth. Oh well, you can always stock your components supply with other goodies you might use later. Or, you can save up your parts needs until you have a big enough order. That'll delay your repair work a long time, though.

Here are some places to look for components:

- **RadioShack (radioshack.com)**   This seller's parts variety is small, but the company offers a few transistors and chips, along with standard 5 percent resistor values and some electrolytics.
- **DigiKey (digikey.com)**   This mail-order parts house has just about everything you could ever want. Its catalog is overwhelming, and you can download it as a PDF.
- **Mouser Electronics (mouser.com)**   Another powerhouse, Mouser has a wide variety of components, including many used in consumer electronics devices.
- **All Electronics (allelectronics.com)**   This is a surplus house with lots of interesting material at bargain prices. It has inexpensive, generic backlight inverter boards that can be retrofitted to LCD monitors, though the boards lack terminals for brightness control.

Do an online search and you'll turn up dozens more sources for both prime and surplus components.

Saving Damaged Boards

When you desolder a through-hole component, one unfortunate result of failing to get the hole hot enough is that its copper lining comes out with the lead. If you see what looks like a sleeve around the lead, you've torn out the copper. On a double-sided board, it's not a catastrophe. When you replace the part, be sure to solder both the top and bottom contact points, and all will be well. You might have to scrape some of the green *solder mask* coating off the top area to get contact between the lead and the foil. That's best done with the tip of an X-Acto knife.

Pulling the sleeve out of a multilayer board can destroy it because you have no way to reconnect with interior foil layers that were in contact with the sleeve. If you're lucky, that particular hole might not have had inner contacts, and soldering to the top and bottom may save the day, so it's worth a try. Don't be surprised, though, if the circuit no longer works.

If you can figure out where they go, broken connections can be jumped with wire. On double-sided boards, it's not too hard to trace the lines visually, though you may have to flip the board over a few times as you follow the path. When you find where a broken trace went, verify continuity with your DMM, from the end back to the break. Don't forget to scrape off the solder mask where you want to contact the broken line.

Wire jumping can help save boards with bad conductive glue interconnects, too. On a double-sided board, you can scrape out the glue and run a strand of bare wire through the hole, soldering it to either side. Forget about trying this on a multilayer board, however; you'll probably trash it while trying to clean the hole. On those, it's best to run an insulated wire around the board from one side to the other. That adds extra length to the conductive path, which could cause problems in some critical circuits, especially those operating at high frequencies. At audio frequencies, it should be fine. If some interior layers are no longer making contact with the glue, this won't work. Most conductive glue boards I've seen have been double-sided, making them suitable for wire jumping.

If the board is cracked from, say, having taken a fall, scrape the ends of the copper lines at the crack. It's possible to simply solder over them, bridging the crack, but that technique tends to be less permanent than placing very fine wire over the break and soldering on either side. To get wire fine enough, look through your stash of parts machines for some small-gauge stranded wire. Skin it, untwist it and remove a single strand.

Sometimes there are multiple broken lines too close to each other for soldering without creating shorts between them. To save boards like that, scrape the solder mask off close to the crack on every other line. Then scrape the in-between lines farther away from the crack. Use the bare wire strands to fix the close set, and use wire-wrap wire (very thin, single-strand, insulated wire used with wire-wrap guns for prototyping experimental circuitry) or enamel-insulated "magnet wire" to jump the farther set. Wire-wrap wire is especially good for this kind of work because its insulation doesn't melt very easily, so it won't crawl up the wire when you solder close to it, exposing bare wire that could short to the repaired lines nearby. Plus, it's thin enough to fit in pretty small spaces. For even tighter environments, use the magnet wire. Just be sure to tin the ends of the wire to remove the enamel, so you'll get a good connection.

It's possible to repair broken ribbon cables in stationary applications (the ribbon doesn't move or flex), if they are the copper-conductor type of ribbons, not the very thin, printed style. Fixing cracks with wire is a tedious, time-consuming technique, but it works. Accomplishing it without causing shorts takes practice and isn't always possible with very small, dense boards and ribbons.

On multilayer boards, cracks and torn sleeves are extremely difficult to bypass. If you have a schematic, you may be able to find the path and jump with wire. Without one, it's pretty much impossible when the tracks are inside the board.

LSI and Other Dirty Words

Back in Chapter 7, I promised to describe a trick for resoldering big ICs with very close lead spacing. Those large-scale integrated (LSI) chips with 100 leads are in just about everything these days. It's not likely you could find a replacement chip, so why would you want to resolder one?

With so many leads, an intermittent connection to an LSI is not uncommon. SMT boards are factory-assembled with *reflow soldering*, in which solder is applied to the pads and then reflowed onto the component leads with hot air, infrared lamps or in a special oven. Reflow soldering relies on low-temperature solder that can break after the numerous heating and cooling cycles encountered in a product's normal use. Now and then one connection out of an LSI's long row of them will go flaky. The leads are so close together that there's no way to apply solder to one without causing a short to the adjacent leads.

Here's the trick: go ahead and short them! With all power removed, of course, solder away and let as many leads get shorted together as you want. Once you have good solder on the problem lead, lay solder wick across the leads where the excess solder is shorting them. Heat it up and wick off the excess, but don't wait until the leads are bone-dry. Pull the wick off a little sooner. If you get the timing right, you'll be left with a perfectly soldered row, with no shorts. The wick soaks up the solder in between leads faster than what's underneath them (where you want it to stay).

If you wait too long and wind up removing so much solder that the connections to the board aren't solid anymore, resolder the area and do the wick trick again. After you try this procedure a few times, you'll get the hang of how long to wait before pulling the wick. I've had tremendous success with this approach. The one caveat is that it's hard to wick out solder if it gets under the edges of the chip. To avoid that problem, solder as far from the body of the IC as possible. That helps prevent overheating the chip, too.

When you're all done, use your magnifier to verify that the contact points with the board are soldered, and that no bridges exist between leads. Honest, it really does work! I've even replaced a few LSIs this way, using chips from parts units. Getting those babies lined up accurately on all four sides...well, that's another story.

Chapter 13

That's a Wrap: Reverse-Order Reassembly

Y̶ou fixed it. Congratulations! Now it's time to put everything back together. Just screw the boards down, plug in the connectors, close the lid and you're done, right? Well, sometimes, but not performing the reassembly methodically can lead to all kinds of trouble, from failure of your repair to new damage, and even to danger for the product's user. You're sewing up your patient after the operation, and it's important to put in the stitches carefully and avoid leaving a scar.

Common Errors

It might seem absurd to think that one could reassemble a machine and have parts left over, but it happens all the time. You snap that final case part into place, breathe a sigh of relief, glance at the back of your workbench, and there it is: some widget you know belongs inside the unit, but you forgot all about it. Hmm, is it really worth all the trouble to backtrack, just for one little, seemingly non-essential item? Sigh.... Time to pull the whole mess apart again. It's very easy to forget to replace a bracket, a washer, a shield, a cover or even a cable. The unit might function without one of those pieces, but it's not going to work completely right.

Putting the wrong screws in the wrong places may have no consequences, but it also could seriously damage the device. If a screw that's too long presses against a circuit board, it might short whatever it touches to the chassis, hence to ground. You flip on the power and, voilà, you have a new repair job on your hands. See the definition of *magic smoke* in the Glossary.

Overtightening screws can strip their heads, making it very hard to get them out again. It can also break plastic assemblies and cause cracks in the case. Undertightening screws may lead to their falling out later, possibly jamming mechanisms or shorting out circuitry if they're internal screws.

Our memories can really fool us sometimes. You're certain that part went over here, but now it doesn't quite fit. So, you press a little harder, because you *know* that's where it goes. Snap! Oh, right, it went over there, not over here. If something doesn't want to fit together, it probably belongs somewhere else. This kind of error happens more with mechanical parts than with electronic components.

Manufacturers try to make internal connectors different enough from each other that only the correct cables will fit. In products with lots of connectors, though, there may be ambiguity. CRT projection TVs, for instance, have so many connectors that there's just no way to make them all distinct, and sometimes a color code is all you have to guide your reassembly. It's not hard to get one wrong. Guess what happens when you turn the set on.

Even when you put ribbon cables in the right connectors, it's easy to damage them. They're delicate and easily torn or folded hard enough to snap their printed copper conductors. The connectors are fairly breakable too, especially the tiny ones in pocket-sized products. Those plastic sliders that hold the cables in place snap off without much pressure. Worse, the entire connector can break its solder joints and fall right off the board, just from the stress of being pushed on while the cable is inserted and clamped down. I've seen that happen on digital cameras and pocket video gear.

Getting Started

To begin reassembly, reverse the order in which you took the machine apart. When the unit has multiple boards, you'll need to get the inner ones reinstalled first. If a board that'll wind up under another board has connectors, put the cables in before covering up that board.

In older gear, take a look at the ground lands on circuit boards, where the screw or the metal bracket makes contact. They exist to connect those points to ground via the chassis, and a poor connection due to oxidation or corrosion can seriously affect the product's performance. Clean them up with some contact spray or, in extreme cases, fine sandpaper. Make sure the screws are tight, so contact will be reliable, but don't overtighten to the point that you might crack the board. Heating and cooling in larger items, and physical stress in portable devices from being bounced around, can cause cracks later on if the screws are extremely tight.

A little sealant on a screw head is better than pushing the limits of tightness. Manufacturers and pro shops use a type of paint called *glyptal* to keep screws from loosening. Swabbed around the edges of the screw head, it is highly effective. You can use nail polish. Don't glob it on; just a little smear will do fine. Be careful not to cover components or their leads, and let it dry before closing the case, so the outgassing won't remain inside. I use red polish so I can see where I've been, should I open the unit again later on.

Placement of wires and cables is called *lead dress*, and it can be surprisingly important. When you took the unit apart, you may have noticed that some wires were tacked down with hot-melt glue or silicone sealer. If the manufacturer went to the trouble to do this, there was a reason. Maybe the wire needed to be kept away from a hot heatsink that could melt its insulation. Perhaps a cable carries a weak, delicate signal that would receive interference if it got too close to some other element of the machine. This can be the case with the cables going to video head drums. Or maybe the reverse is true: the cable would cause interference to other sensitive circuits. Wires carrying high voltages, like those used to run projector lamps and LCD backlights, may need to be kept away from all other circuitry to avoid not only interference but the possibility of arcing. The closer a high-voltage wire is to ground, the more those devious electrons want to punch through the insulation and get there. Give 'em time and they just might.

If the manufacturer tacked wiring down, put it back the way you found it. Hot-melt glue is somewhat flammable, and it melts with heat, of course, so it isn't used much in larger products. Now and then, you may find it in smaller items that don't carry much voltage or produce significant heat. To tack wires back down into it, you can melt the blob with your plastic-melting iron, avoiding any other wires, or drip a little more glue on top from your glue gun.

More often, you'll find silicone sealer used to secure wires. The type used is called RTV, and it's best to replace it with the same kind, because it offers the correct insulating strength. RTV is available at most hardware stores, and electronics supply houses carry it too.

Now and then, and especially in small-signal RF stages operating at very high frequencies, you'll see blobs of wax covering transformers and capacitors. The wax holds the parts to the board and dampens vibrations that can cause noises in signals or frequency instability in oscillators. In some circuits, the capacitance of the wax may affect circuit operation, so it's best to remelt and reuse the original wax. It's not the same stuff that's in the candles on your dinner table.

Even if the manufacturer took no extra care with wiring, you should pay attention to the issue to achieve maximum product reliability and safety. Could a power supply lead touch a heatsink? Is the cable from a tape head going right by the power supply? And, perhaps most important, is any wire or cable placed such that it'll get crimped by a circuit board or part of the case when reassembly is complete? The sharp ends of a board's component leads can go right through a wire's insulation, with disastrous results. Crimping caused by the case can break the wire or cut through the insulation and short it to the chassis. With a wire carrying unisolated AC power, a crimp could even present a shock hazard. If the case halves don't mate properly, a wire is probably in the way. Don't just squish them together and go on.

These scenarios may sound farfetched, but they're really pretty common. I always make it a point to watch the wiring as I close up the case, imagining where things will be and how they'll press on each other before I actually snap the halves together or tighten the final screws.

Reconnecting Ribbons

Insert ribbon cables into their sockets carefully. It isn't hard to put them in wrong, which can lead to anything from no operation to circuit damage. Most ribbons have bare conductor fingers only on one side. If you get one of those in upside-down, the product won't work, but it's unlikely to cause damage unless the socket has U-shaped contacts that touch both sides of the ribbon. There are some like that.

Double-sided ribbons offer more opportunity for calamity. How do you know when they're in the right way? Many are keyed with a notch at one end so they can't be inserted upside-down. Some are not, though. I sure hope you heeded my advice in Chapter 9 to mark the darned things! If not, see if the cable has a bend or curvature suggesting its original orientation.

As discussed in Chapter 9, some ribbon connectors have no latches, and the ribbon just slides in. That type requires some force for proper insertion. With such a connection style, the ribbon cable will have a stiff reinforcement layer at the end. Even when you find one, take a good look at the connector to be sure it has no latch that slides in or flips up, because some latch-type connectors accept reinforced ribbons too. If you see no latch, grasp the cable's reinforced tab, carefully line up the ribbon with the connector and press it in firmly.

Connectors with latches are easier to manipulate. First, be sure the latch is open. Slide it out or flip it up. The flip-up kind will stay up while you insert the ribbon, but the slide style has an annoying habit of going partway in before you want it to, blocking full reinsertion. One end may slide in, resulting in a crooked latch. If that happens, pull that end back out, remove the ribbon and try again. You might have to hold the slider's ends with one hand while you insert the ribbon with the other.

Latch-style connectors require almost no insertion force. Gently slide in the ribbon until it stops. Don't press firmly here. Look down at the top of the connector and verify that the ribbon isn't crooked. Then close the latch while holding the ribbon in place with your other hand. That's easy with flip-up latches and a little harder with sliders. Occasionally, I've had to close sliders one end at a time with a thumbnail or a screwdriver while holding the ribbon in my other hand. To avoid a crooked result, it's better to close the ends at the same time, but now and then you gotta do what you gotta do.

When the latch is closed, look again at the exit point and make sure the ribbon is straight. You may see the edges of the bare conductors sticking out. That's fine as long as they're even and you're sure the ribbon is in all the way. Many of them are designed that way.

Special ribbons used for hard drives and other very dense applications can have two sets of fingers, one behind the other. These will always have latch-style connectors. Be absolutely certain the ribbon is fully inserted, so there's no chance the wrong set of fingers could make contact with the mating pins in the connector.

Oops!

If you were unlucky enough to break the latch on a sliding-style connector when you removed the ribbon, don't despair. The object is to get pressure on the conductive fingers so they make good connection with the socket. Find some thin, soft plastic from, say, the bottom of one of those little pudding cups in which you keep screws.

Cut the plastic into a rectangle that just fits into the socket and sticks out a few millimeters. Trim carefully so the edges line up well with the edges of the socket, without a gap. Now put the ribbon in and wedge the plastic piece in to replace the broken latch. Be absolutely sure to insert it on the side of the ribbon that does *not* have the conductive fingers, or the plastic will block the connection. I got that wrong once and went around in circles for hours trying to figure out why that confounded shortwave radio wouldn't turn on anymore!

If you get the thickness right, it'll take a little pressure to slide in the plastic, but not a lot. If it slides in very easily, it may not put enough pressure on the ribbon to make good connections. The few times I've had to do this, I used forceps to push in the plastic piece. Needlenose pliers will work as well.

Flip-up latches are much harder to repair. It might be possible to modify the socket by melting a piece of plastic over it and converting it to a sliding arrangement, and then using the plastic insert approach, but it'd be a difficult modification to pull off, considering the size scale of some of these connectors. Trivial as it may seem, a broken flip-up latch often means the end of the product unless you can scrounge a latch from a parts unit.

Layers and Cups

Here's where those pudding cups come into play. If you used them as I suggested back in Chapter 9, your innermost layer's screws will be in the cup second to the top, just under the empty protective cup. Put that layer's boards, shields and assemblies into place and fasten them with those screws. If the screws are of different sizes or styles, you should have taken digital photos or made a drawing and placed it in the cup with the screws. Be certain to use all the loose parts in the cup, because you won't be able to reach that layer once you reinstall the layers covering it.

When you're ready for the next layer, pull the empty cup and put it aside, exposing the next set of screws. Continue on with the next layer, and so on. When you're all done, you should have a nice set of empty cups ready for the next project. If you have screws left over in the final cup, remember to check under labels and rubber feet for hidden holes.

Oh, Snap!

Those nasty hidden snaps are much easier to close than they were to open! Line them up carefully and apply pressure until they pop into place. The edges of the case should fit smoothly. If there's a bulge, either the snap isn't all the way in or a wire is caught underneath.

Sometimes a case has to be snapped together at one end before the other, even if it didn't come apart that way. Look at the style of snap, and how it fits together should be apparent. If you get it wrong, you might break a snap, but it's not a big deal. Heck, they can break even when everything is done right. Often you can live without one or two. If a snap's loss makes the case wobbly, it might be worth some careful repair with your plastic-melting iron. Be sure to pop the case apart first; trying to melt plastic near the outside will almost certainly result in very visible damage. Even from a half-inch away, the iron puts out enough heat to soften and deform many plastics. A repaired snap is weak, so you may get only one chance to close up the case properly. Still, it's better than nothing. If you hear something floating around inside the unit after you finish putting it together, the snap has broken off. Open the case and remove the plastic piece.

Screwing It Up Without Screwing It Up

As I mentioned, screws should be reinstalled carefully to avoid damaging them or the plastic into which they are screwed. Phillips screw heads are especially easy to strip, and trying to remove one is mighty frustrating once you do. Insert all the screws in a layer, or on the outside of the case, but don't tighten them down before the last one in that layer is in its hole. Sometimes you'll need to remove one because it's the wrong length or you suddenly realize the black one went here and the silver one over there. Or, an internal bracket doesn't quite line up and you need to open the case again before going on. Once you have them all in, it's time for the final tightening. How tight is right? Hold the screwdriver with your fingers, not in your palm. Turn the screws just until they stop, and snug them in ever so slightly. That's it. Don't twist until you can't twist any farther.

Done!

If everything fits together well, you should be ready to fire up the unit and consider your repair complete. Be sure to bench-test receivers, projectors and other heat-generating products if the work you did could possibly make them run too hot. You'll want to bench-test a projector whose fan or ballast circuit you repaired or replaced, or a receiver that needed new output transistors. A digital camera or an MP3 player, of course, won't require that extra step. Now go show off your work and bask in the glory of a job well done. You've earned it!

Chapter 14

Aces Up Your Sleeve: Tips and Tricks for Specific Products

Although the principles we've covered apply to pretty much all electronic devices, various product categories are different enough that they benefit from specific troubleshooting techniques. Let's look at some of the most common gadgets, their typical problems, and how to approach their repair.

Switching Power Supplies

Since you'll run into issues with switching power supplies in many kinds of machines, let's cover them first, before looking at the products in which they take up residence.

How They Work

Switching supplies rectify the incoming AC into DC and then chop it at high frequency, pulling current through a transformer's primary coil with each pulse. Using a high frequency allows the energy to be replenished on the other side of the transformer much more frequently than with the old linear approach, which was limited to the 60-Hz line frequency. So, the transformer doesn't have to convert as much power at one time and can be a lot smaller. The approach also keeps the chopper transistor either saturated (turned all the way on) or cut off (turned all the way off) most of the time, resulting in high efficiency, since it spends almost no time per pulse in its midrange (partially turned on), where transistors act like variable resistors, with all the attendant heat resistance generates.

What Can Go Wrong

Controlling those high-frequency pulses, with their fast rise and fall times, is tougher and stresses the components a lot more than did low-speed linear circuits. The fast pulses with high-voltage peaks punch holes in transistors' substrates, and the rapid charging and discharging of filter capacitors wears those out too. Consequently, switchers fail significantly more often than do linear supplies. Nonetheless, very few products still use the old technology; switchers are everywhere, from computer supplies to little AC adapters and chargers for cameras and cell phones.

Is It Worth It?

Unless the transformer is shorted or open, it's generally worth repairing a switcher, especially if it's in a product you want to get working again. It might be a waste of effort on an AC adapter that could be easily replaced. The transformer rarely goes bad, though I've seen it happen now and then with that ubiquitous cousin of the switching supply, the backlight inverter. The high voltage of an inverter's output sometimes breaks down the insulation between the transformer's windings. In a normal, AC-powered switcher used to create low-voltage DC output, that's an unlikely scenario.

If multiple semiconductors have blown from a chain reaction feeding voltage from one stage to the next, fixing the supply may be more trouble than it's worth. Especially if the pulse-width modulator chip is dead, getting the part can be a hassle.

The Dangers Within

To service a switcher, disconnect the power first! *Never* work on one with power applied unless you have an isolation transformer. Even then, don't work on powered switchers until you gain a fair amount of experience. The dangers are real and significant. Did you take off your wristwatch and all jewelry? It's especially important now. Be sure to wear shoes, too.

Switchers are organized in two sides: the primary side, connected to the AC line and with no ground connection to the rest of the product's circuitry, and the secondary side, isolated from the AC line by the transformer and connected to circuit ground in the rest of the device. Linear supplies share the same basic organization, except that there's no circuitry on the primary side beyond a fuse.

The primary side of a switcher is where most of the trouble is, and also most of the danger. The chopper circuit stresses its transistor harder than any other component in the supply, leading to frequent failures. The side's direct connection to the house wiring makes service hazardous because any contact between you and ground completes the circuit. The secondary side is at lower voltages, and its isolation from the line means it's a lot safer. Some switchers go as far as having holes in the circuit board between sides for some extra protection from arcing over and loss of isolation.

How to Fix One

Most switchers fail from bad electrolytic capacitors, blown rectifiers or a dead chopper transistor. Look at the capacitors first. Any bulges? Change them. Leakage? Change them. Anything at all unusual about their appearance? Change them!

Checking the rectifiers is easy enough if they're separate diodes. When you have a bridge rectifier, with all four diodes in one package, each diode must be tested as if it were a separate part. Take a look at the bridge rectifier diagram back in Chapter 7. With all power disconnected and the big electrolytic near the bridge discharged, desolder the bridge from the board and use your DMM's diode function to test each diode in it. You should see around 0.7 volts drop at each diode in the forward direction and an open circuit in the reverse direction, as with any silicon diode. If you find an open or a short in any of the diodes, replace the bridge.

The chopper is the big transistor, probably heatsinked, on the primary side of the transformer. Some choppers are bipolar transistors, but most are power MOSFETs. If the fuse is blown, it's a good bet the chopper has shorted out. The transistor can fail open, too, in which case the fuse might still be good. The transistor may have shorted and then opened, and the fuse may or may not have survived the momentary overcurrent. It's an old technician's anecdote that transistors are there to protect fuses! Check the chopper using the out-of-circuit techniques discussed in Chapter 7.

If you have an isolation transformer, you can do some powered tests before pulling parts. Check the voltage across the big cap on the primary side of the supply, near the chopper. Remember that you can't use circuit ground on this side. The negative terminal of the cap will be your reference point, where you'll connect the meter's black lead. You should see at least 300 volts. If it's much less, suspect a bad bridge rectifier. If it's zero, the fuse is probably blown, which could mean a bad bridge, a shorted cap or a bad chopper.

It's best not to try to scope the chopper directly, as the voltages are very high. The safer approach is to scope the secondary side of the transformer, using normal circuit ground. Many switchers have multiple taps on the secondary winding. Any of them will do, as long as it's not the one connected to circuit ground. If the chopper is running, you'll see pulses at a significantly lower voltage than what's on the other side. They won't be tiny, though. Expect anything from 10 to perhaps 60 volts from the baseline to the peak. No pulses? She ain't running.

If the chopper is good but isn't running, suspect the pulse-width modulator (PWM) chip or the regulation circuitry near the output. Open zener diodes on the secondary side can allow the output voltage to rise too high, activating protection circuitry and shutting down the PWM, or even tripping the crowbar, deliberately blowing the fuse. No pulses, no chopper, no operation.

If the supply is running but not putting out proper power, caps on the secondary side are the primary suspects. Scope them. If you see much of anything but DC on an electrolytic that has one lead going to ground, change it. Either its capacitance has declined, its ESR has risen, or both. If you change the cap but the waveform still looks noisy, look for a leaky diode feeding the cap.

Finally, remember that most switchers will shut down if output current demand exceeds their safe limits. Some may blow their fuses for the same reason. A short somewhere else in the machine may be pulling too much current and causing the supply to act like it's broken.

Audio Amplifiers and Receivers

Audio amps and receivers are at the centers of all home theater setups. The units have to produce significant power to drive speakers, so they include a fair amount of heat-generating circuitry prone to failure.

How They Work

Though today's audio amplifiers and receivers employ digital signal processing for surround sound decoding, delay effects and sophisticated tone controls, power amplification is still an analog process in nearly all of them. The exception is the Class D digital amplifier, which converts the incoming signals to pulse-width modulation of a high-frequency carrier. The pulses are then current-amplified, with a lot of amperes available in each pulse. At the output, a smoothing filter blends them back into audio before sending them to the speaker. Class D amplifiers are used mostly in automotive applications because they are extremely efficient and can develop a lot of power in a small box without getting very hot. Fidelity can be quite high in Class D, but achieving it isn't easy. So, home audio gear, which lives in a quiet environment more conducive to critical listening, has stuck with analog amplification, a very mature, refined technology capable of exceptionally good sound reproduction.

Before home theater, the chain was simple: preamp, through tone controls, to drivers and outputs. Not anymore! Input may come from analog jacks, digital coaxial or digital optical cables, with varying sample rates and bit depths (number of bits per sample). Several formats of multichannel encoding are used, too, and the unit has to be able to handle them all. Once the desired signal processing has been accomplished, the data is converted back to analog and applied to conventional audio circuits. Capacitive coupling, with each stage connected to the previous and next stages via capacitors, is rarely used because the caps cause phase shift and rolling off at the low end of the audio frequency spectrum. Most of today's amplifiers are directly or resistively coupled, for maximum fidelity.

Because DVDs brought audio with more than two channels to the home, the conventional stereo receiver is all but gone; newer units have at least five channels: two front, two rear and one subwoofer for deep bass. Some have seven. What's the difference between and woofer and a subwoofer? One of them can operate under water! No, seriously, a subwoofer is for reproduction of only the lowest frequencies, usually under 100 Hz, while a woofer's range may extend into the hundreds of hertz. With today's small "satellite" speakers unable to reproduce low frequencies much at all, the subwoofer is really a woofer, but the term has stuck. One defining characteristic is that there's only one of them, as opposed to the usual separate woofer for each channel, because very low audio frequencies have little directionality, filling the room regardless of the speaker's position. Thus, there's no separation, stereo or otherwise, and the sense of spatiality comes from the higher frequencies being reproduced by the satellites.

For every channel there is a complete amplifier chain culminating in an output stage. Many receivers use power amplifier modules for their outputs, but higher-end units still go with discrete stages because they're reputed to sound better.

What Can Go Wrong

The power supply works mighty hard, at least when the volume is turned up high. Moving a lot of air with the subwoofer, necessary for the serious bass frequencies found in movie explosions and such, takes an especially large amount of power. While modest home systems may have a hundred watts available for that, self-amplified subwoofers with several *thousand* watts have been marketed. Of course, people utilizing the full power of those things can't actually *hear* their movies anymore, but that's what subtitles are for, right?

Plenty of receivers still use linear power supplies because switchers can introduce high-frequency noise that's hard to eliminate. To power five channels of 100 watts each takes some serious iron in the transformer, along with high-current diodes, hot-running linear voltage regulators and huge electrolytic capacitors. There's a reason receivers weigh so much! All that heat and high current take their toll, especially on the diodes and capacitors. Even with the generally higher reliability of the linear approach, power supply failures in receivers are common.

The most trouble-prone parts of a receiver are the output stages. Whether modules or separate transistors, they are where the current is. Plus, they operate in their linear regions, neither saturated nor cut off at any time (one hopes!), so they're essentially resistors, dissipating power supply current as heat. Look for large heatsinks and you'll find the output stages.

With all those jacks, input switches and interconnections between boards, signals can be impeded by bad connections, causing them to crackle or drop out completely. Phono preamp sections are especially vulnerable to this, as they handle the very tiny signals generated by magnetic phono cartridges, and it doesn't take much to stop those. Cartridges put out around 5 mv peak-to-peak, compared to the 1-volt standard for line-level audio. Still, even with the higher-level signals, ratty connections in the signal path cause many receiver problems.

Speaker protection circuits sense when there is significant *DC offset*, or variance at the midline of the signal waveform from 0 volts. When offset occurs, there's a fault somewhere in the amplifier, and the protection circuits disconnect the speakers to prevent excessive power supply current from burning out their voice coils. At least that's how they're supposed to work. Now and then the protection circuit malfunctions, going into protection mode when nothing is really wrong.

Is It Worth It?

Not much in a receiver's power supply or amplifier chain is especially expensive or hard to get, so most receivers of any real value are worth repairing. Output modules are available from online parts houses, as are power transistors. You aren't going to find the DSP (digital signal processing) chips, but it's highly unlikely you'll need them anyway. In receivers, the power-handling areas are usually where the mischief lurks.

The Dangers Within

Most output stages are complementary push-pull types fed by positive and negative power supply lines. (A push-pull design with only one power supply polarity is called *quasi-complementary*.) The supply lines may have 30 to 80 volts or so on them, so they are capable of shocking you. And should you be unlucky enough to touch both at the same time, you could come in contact with 160 volts or more.

Receivers with fluorescent display panels have small inverters supplying the panels with the few hundred volts required to light them up. The inverters are typically located on boards just behind the displays.

Heatsinks can get mighty hot, so avoid touching them if the unit has been on for awhile, especially at high volume.

How to Fix One

Servicing a receiver is very much a process of elimination. The first thing to consider is whether the problem affects all the channels. If so, head for the power supply. One channel could have a short pulling everything down, but the supply is the first thing to check. If only one channel is affected, that's where you'll find the trouble; the supply has to be okay, since it's powering the other channels properly.

Power Supply Problems

With a linear supply, testing is pretty painless, compared to the ordeals of working on a switcher. The transformer comes early in the chain, and what's between it and the AC line is easy to evaluate without power applied. If the unit is dead, check the fuse. As with other products we've examined, a blown fuse almost always means something pulled too much current through it. Look for shorted rectifiers or a short somewhere farther down the line, on the other side of the transformer. If there's a connector you can pull to isolate the supply from the rest of the unit, try doing that to see if the overcurrent situation persists.

Output Stage Problems

A shorted output stage will pull the supply's output down toward ground, possibly blowing the fuse. Sometimes there's enough resistance between the shorted transistor or module and the supply to limit the current to a value below the fuse's rating, so it survives. The telltale sign of a shorted output stage is a loud hum through a speaker connected to that stage. However, other channels' outputs may also produce the hum because they're fed from the same supply line that's being pulled down by the bad one. They won't hum as loudly, though. Scope the output. If you see DC with something resembling a 60-Hz sawtooth wave riding on it, you've got a short.

Frequently, changing the output transistors or the module takes care of everything. Sometimes, and particularly with discrete designs (no module), the driver transistors may be shorted too. Also, always check the resistors in series with the emitters of the power transistors in discrete stages. The overcurrent can really cook those babies, altering their values or even cracking them. Most techs replace the emitter resistors as a matter of course when they change output transistors.

In direct-coupled designs, shorts in one stage may blow surrounding stages. I've seen a shorted output transistor take out every transistor in its channel, right back to the input jacks! That used to happen frequently in early designs, but it's still possible even now. Ironically, the costliest units are more likely to have extensive damage, thanks to their closely coupled stages with few or no components to isolate them. That kind of circuit sounds the best when it works, but it's a mess when it breaks.

At the inputs to the driver stage, look for diodes or zeners. They're used to set the bias points, keeping the positive and negative halves of the output stage just slightly turned on even when there's no signal or it's very small. An open zener or a shorted diode can make the bias go wild, causing the outputs to conduct themselves to death. Changing the transistors without checking the bias just means you'll need more transistors in a few minutes.

The damage can proceed the other way, too, from input to output. Again, designs with close coupling are the most susceptible because their output stages' bias is set right at the start. If the input transistor shorts, it can drive DC right to the outputs, upsetting their bias and sending them to semiconductor heaven. Or, considering the heat, perhaps it's the other place.

Although it seems like diagnosing an amplifier should be very straightforward, the interdependence of DC levels from stage to stage can turn a romp into a nightmare. If you find yourself going around in circles, disconnect the output stages from the power supply and scope their input lines. You should see normal audio with some DC offset. Most receivers use bipolar output transistors, so the bias is a matter of base *current*, not voltage. With no power to the transistors, the offset may be significant, but it shouldn't be close to the power supply rails.

Small-Signal Problems

Small-signal issues are like those in any other device. You could have a bad cap, a bad transistor, and so on. To determine whether the problem is in the small-signal sections, test the various inputs. If *any* of them works properly, the power supply and output stages are doing their jobs. The various signals—digital, analog, radio—wind up as analog audio at the inputs to the power amplifier stages, starting with the line-level (1 volt peak-to-peak) stage. If no input works, go back to the beginning of the signal path, using the simplest one possible: an auxiliary analog jack. Feed it a signal from an audio player, select it on the front panel as the active audio source, and scope your way through the low-level stages to see where it stops. Oh, and be sure to turn the "tape monitor" switches off! They interrupt the audio path so a tape deck can be inserted in line with it. With no deck connected, the audio comes to a dead end, never reaching the power amplifiers.

With their multiple input paths, receivers have more connections through switches than most devices, offering plenty of opportunities for bad connections. Newer units often have electronic switching, rather than the old rotary or pushbutton switches used for decades. Noncontact switching is more reliable, but a blown analog switch chip will stop things dead. It's pretty rare for just one of the chip's inputs to fail, though; a bad chip usually kills them all.

Hum isn't always a power supply or output stage problem. If the sound of the hum is thin, with high-frequency content that's a little bit buzzy, and it isn't loud enough to wipe out the audio, there may be a bad ground connection somewhere in the low-level circuitry. As in many products, grounding from the board to the chassis can be mechanical, provided by the pressure between the board's ground lands and the metal tabs into which the board is screwed. Over time, oxidation increases the resistance of those connections, and hum can result.

If the level of the hum varies with the volume control, the source has to be early in the chain, before the control. Try unplugging the audio source from the input jacks. If the hum disappears, you have a bad cable feeding the jacks or a *ground loop* between the audio source and the receiver. If it's still there, pull the board and clean the ground lands and the chassis mounting tabs until they're nice and shiny. If that doesn't solve it, check for bad solder joints that may be causing poor grounding.

Disc Players and Recorders

Though inexpensive, CD and DVD players are not simple. They are actually little computers, combined with the mechanical and optical sections necessary to retrieve data from the disc.

How They Work

To find, follow and decode the disc's microscopic optical tracks requires a focused laser beam, a three-axis servo system and a fair amount of computing power. An awful lot has to go right for the data to be read from the disc and transformed into your favorite movie or music.

First, the disc must be accepted into the player, properly seated on the spindle and clamped down. The machine moves the optical head to the center of the disc, where playback will begin. Then the laser turns on and the lens moves up and down while the phototransistors in the head look for a reflected beam from the disc. If no reflection is detected, the player stops and displays "no disc." Once it sees a reflection, the machine stops the lens when proper focus is found. This is determined by reception of maximum beam strength in the head's center detector with minimum beam strength in the side sensors, indicating optimum spot size. Only if focus is achieved will the player spin the disc and try to read the track. If the lead-in track is found, the sled motor starts moving the head away from the disc's center, and playback begins. As the disc spins, its normal wobbles and eccentricities far exceed the size of its tracks, so the tracking and focus servos keep the lens dancing around in three-dimensional step with their movements. The disc's speed gradually decreases as the head proceeds toward the outer rim, because the linear distance for each rotation increases. The object of the technique, called *constant linear velocity*, is to keep the speed of the track constant as it goes past the head, so the bits can be crammed in as tightly as possible for maximum disc storage capacity. The spindle speed is controlled by the microprocessor's monitoring the rate at which data is being read from the disc. Only when all these systems work in concert can a disc be stably tracked and played.

What Can Go Wrong

The disc may seat incorrectly, resulting in more wobble than the focus and tracking servos can handle. Many tray-loading players drive the tray with a belt that stretches over time, preventing the door from opening or closing fully. They use nylon gears that can crack and stick whenever another gear's mating tooth hits the crack. The leaf switch telling the micro when the door is open or closed can bend or become oxidized, so the door motor keeps running even after the door hits its limits.

Portable players with top-loading lids have their own quirks. To prevent you from looking directly into a running laser, they use a small interlock switch to sense when the door is closed. A bad switch makes the micro think the door is open and results in no operation. This is a very common failure in these machines.

The sensor indicating when the head is at the starting position can malfunction. Most players use a leaf switch, though some use an optical sensor. The optical variety is pretty reliable, but leaf switches get oxidized or corroded and stop passing current, so the microprocessor never gets the message that the head is positioned. The result is a clacking sound made by the head as it slams over and over into the end of the sled's track.

Unlike LEDs, laser diodes have finite life spans, and they get dimmer as they age. If the laser is too dim, the reflected beams will be hard to detect and decode properly, and the machine will have trouble starting discs. It may also skip on discs that do play, although there are other causes of skipping. And even bright lasers can develop odd internal reflection modes resulting in optical impurity; the beam stays bright but can't be properly focused.

If the ribbon cable's connections to the head are oxidized, the weak signals from the photodetectors will be erratic, confusing the machine badly and causing symptoms much like those of a failing laser.

If the sled motor has a flat spot on its commutator (where the brushes transmit power to the rotating coils), or the slide or gears need lubrication, the head will stick as it scans across the disc, resulting in skipping or freezing. It's important to remember that the sled motor does not move the head in the tiny increments required to advance it along with the moving track; such fine mechanical motion would be impossible to achieve in an affordable product, if at all. Instead, the *lens* moves sideways, even as it bobs up and down to maintain focus, until it approaches the point where the beam would miss the sensors. The micro detects when the lens is near its limits and pulses the sled motor, advancing the head. The lens then moves back to the center and the process begins anew. Essentially, the lens and the sled play a game of inchworm as they follow the recorded track across the disc. So, intermittent, tiny jerks of the sled motor are completely normal. If you observe the sled assembly of a working player, you'll see the wormgear shaft twitch every few seconds as it pushes the head outward ever so slightly.

If the spindle motor, which spins the disc, is worn out or gummed up, the disc may not come up to proper speed, resulting in a slow data rate the machine can't process into normal audio or video content.

And, of course, all the servo and decoding circuitry has to work properly. Playing an optical disc is a feedback process; the data rate tells the spindle motor how fast to turn, and the reflections from the disc surface tell the tracking and focus servos how to keep a grip on the track. A loss of any of these systems can keep the others from doing their jobs.

Is It Worth It?

If the laser is dead, forget about repairing the machine. The alignment of the laser diode in the head is critical, so you can't pull the laser and replace it; you need a whole new head. When disc players were costly, replacement heads were available from parts houses. These days, the players are so cheap that there's no market for the heads, so their sources have dried up.

By the same token, if mechanical parts like motors or gears are broken, your only hope is a parts machine; you're not going to find replacements.

Disc players incorporated into game consoles are the notable exception. Game units still cost enough that heads and even entire sled assemblies are available. Check the Internet for parts houses supplying these items.

Just about anything else can be fixed. You're not going to find a source for large-scale-integrated (LSI) chips and such, but those are unlikely to be causing the trouble anyway, and you couldn't change them even if you tried.

The Dangers Within

Never look directly into the laser beam! The wavelength used for reading CDs is in the infrared, but the purity isn't perfect, so there is some visible red as well. The red portion of the output is dim compared to the primary infrared energy, so looking at it gives a false sense of what your eye is receiving. It's like staring at a solar eclipse: your poor retina is getting blasted but you don't know it. DVDs are played with a visible red laser. It's just as damaging, but at least you can see what's coming at you.

In either case, you might burn a permanent hole in your visual field if you look straight down the bore.

It's routine to check for a working laser, but always look from off to the side. There's enough reflection in the lens to let you see if the beam is there. When you see how bright a DVD player's beam is, even from the side, you'll get a good sense of just how damaging a full-on view from either a DVD or a CD player could be.

Under no circumstances should you ever try to view the laser of a disc *recorder* when it's in record mode, even from off to the side! The optical energy output is high enough to pop balloons; imagine what even a momentary reflection might do to your eyes.

How to Fix One

The most common problems are failure to accept and seat the disc properly, inability to play at all, and skipping while playing. If the door won't open, look at the display. Is the digital control system working? Do normal numbers or messages appear on the display? If not, head for the power supply and check all the usual things like capacitors and output voltages. If the supply is good, scope the clock crystal at the micro. Players typically have multiple crystals, but you should find one right next to the biggest chip on the board. See if it's running. It should have a sine or square wave on it of at least a few volts, at or very near the frequency marked on the crystal.

Door Problems

Assuming the power and control systems are working, check for leaf switch problems around the door. The exact layout varies from machine to machine, but the door motor, gears and belt are usually located under the platform (the entire mechanism), and the leaf switch will be buried in there someplace, with wires going to it.

Reaching the door mechanism requires removing the platform from the frame. Look for large-ish screws at the corners, seen from above. You'll probably have to remove the front panel as well, although sometimes you can pull the platform toward the back and lift it out without doing so. See Figure 14-1.

Some players use two leaf switches, one for the fully out position and one for fully in, but most use one single-pole, double-throw (SPDT) leaf. The center blade gets pressed against one outer blade when the door is open and against the other when it's closed. Look for three wires. Check that the blades aren't bent, and that their contact points are clean. If the contacts are black, their lubrication may have dried out and become insulative. Gentle application of fine sandpaper will take off the black coating.

If the door opens and closes but not all the way, either the belt is shot or there's a broken nylon gear. Nylon gears crack very easily. Any sort of abuse can break them, and sometimes they just fail with time. If the door is very sluggish, the belt is the likely culprit. If it moves at normal speed but stops abruptly at a consistent spot, look for a broken gear. Check between the teeth; sometimes a grain of sand gets in there and jams the gears without breaking anything.

FIGURE 14-1 Door mechanism on underside of platform

Clamp and Spindle Problems

A working door mechanism should take the disc in, plop it on the spindle and then lower the clamp onto it. Believe it or not, the only thing keeping the disc in contact with the spindle is friction. The clamp contains a magnet that's attracted by the metallic spindle, holding the disc in position and facilitating the friction grip.

Portable players rarely have clamps. Instead, three or four spring-loaded ball bearings or tiny tabs in the spindle grab and hold the upper edge of the disc after it is pressed down firmly. I've never seen one of those be a problem.

In door-type players, improper disc clamping is common. The clamp's upper portion hangs loosely in its holder and should float when the disc is spinning. If the holder is misaligned, the clamp will rub against it, making an obvious noise and dragging the disc speed down, perhaps preventing playback. Don't be surprised if the clamp wobbles at the top a bit without rubbing. That's normal. Rubbing, of course, is not.

Dirt on the spindle can keep the disc from sitting flat, leading to focusing errors and skipping. If it gets greasy, the disc may slip. Make sure the spindle is clean.

When the disc spins, it shouldn't wobble much. You may see some shimmering of the surface, but the edges should sit pretty flat. If it wobbles, try another disc, to be sure the first one isn't warped. Assuming a good disc, any significant eccentricity is caused by either a bent spindle shaft (unlikely in door-type players but more plausible in top-loading portables, because people have to press on it when inserting discs) or clamping problems.

Playback Problems

If the disc is properly seated but won't spin, the startup sequence of head positioning, disc detection and focusing has failed. In a door-type player, it should begin as soon as the disc is seated. In a portable unit, there's another element: the interlock switch. When the lid is closed, a little plastic finger on it protrudes through a hole in the player's body, pressing on the switch. If the finger breaks off or the switch fails, the micro will never know the door has been closed, and nothing will happen. I've seen numerous bad interlock switches in these players.

Is the head all the way in toward the center of the disc? If not, there's a sled problem. Check that the rails are greased at least a little, and that the start position limit switch is okay. Sometimes the leaf switch gets bent just a tad and will still work, but not until the head is closer to the spindle than it should be. Focus will be achieved but the head won't find the lead-in track, and it'll just sit there until the machine gives up and stops.

Before going further, check that the lens is clean and not scratched. Portable players with exposed heads are especially subject to dirty and damaged lenses. The lenses in door-type machines are well protected from scratches, but they can still accumulate dirt. If the machine has been in a smoker's home or was installed in a kitchen, there may be a film on the lens, preventing proper focus. Even a protected lens could be scratched if the user tried one of those nasty cleaning CDs with brushes.

Getting to the lens in a portable is easy. Pop open the lid and there it is. Take a cotton swab and wet it with water. Do *not* use alcohol! The lenses are plastic and will be destroyed by it. Blot most of the water from the swab with a tissue, so that it's damp but will not drip water into the optical head. Wipe the lens gently and then wipe it again with a dry swab. Don't put pressure on the lens while doing this.

To clean the lens in a door-style player, you may have to disassemble the clamp assembly and remove it. Sometimes you can reach under it by bending a swab's head at an angle.

With the head at the starting position, the lens should move up and down as the player searches for proper focus. The lens is mounted on coils of very fine wire called *voice coils*, so named because the arrangement of a coil over a magnet is similar to what's found in a speaker. Current passing through the coils generates a magnetic field that interacts with a permanent magnet in the head, permitting the control system to move the lens toward or away from the spindle and up and down. During playback, the lens assembly floats on this field, bobbing and weaving as necessary to follow the disc's track. Look at the head from the edge of the disc and you'll see the lens. If it's not moving, there's a problem with the focus circuits or the cable to the head. Even with a dead laser, the lens should move a few times until the micro figures out there's no reflected beam.

If the lens doesn't move, the start-up sequence has not been initiated. There could be a digital control problem, or the door's leaf switch might not be signaling the micro that the door has been closed. The door motor may still be straining, but you might not be aware of it. In portables, check that interlock switch! If it doesn't tell the micro the door is closed, nothing will start.

Laser Problems

Is there a beam? Its side reflection will be red (except in a Blu-Ray player, where it'll be blue). In a CD player, the beam will look dim. In a DVD player, it's very bright. If it's there, the laser is probably okay. It might still have problems, but at least you know it's not dead.

If there is no beam, you've hit on the trouble! Alas, most disc player failures are due to a bad laser. If the lens is moving but there's no light, the laser is probably shot, and you have just become the proud owner of a parts machine. To be sure, you can trace back from the diode and see if it's getting voltage. Laser diodes are driven by a few volts DC, and there's usually a tiny trimpot right on the head at the diode that'll help you find the right connections. If the DC is there but the beam is not, it's bye-bye laser and bye-bye player.

You may see two trimpots in DVD players. Many use two lasers, an infrared for playing CDs and a visible red for playing DVDs. If the machine will play one but not the other, one of the lasers may have died. Be sure to test the unit with the type of disc it won't play.

Even if there is a beam, it might be too dim for proper operation. It's hard to tell with a CD player, because most of the energy isn't visible anyway, but you can get a good idea by comparing the brightness to that of a good player. In a DVD player, the beam is so bright that often you can see it right through the disc!

If the beam shines and the lens moves, but nothing else happens, it may not be finding focus. Look on the board for a test point labeled "FOK" (yeah, I know), "FOC OK" or "FOCUS OK." Scope it. It'll change state (usually from low to high) when focus is achieved. If it doesn't, then something is preventing proper focusing.

If the disc starts spinning, focus lock is good. The rotational speed depends on where on the disc the head is positioned. At the start, the disc should spin a few hundred RPM. If not, there's a problem with the spindle motor. Either it is mechanically gummed up, the motor is bad, or the driving circuitry isn't doing its job.

Hair, both pet and human, can wind itself around the spindle motor's shaft, even in the protected environment of a door-style player. Remove the disc and turn the spindle by hand. It should turn easily and smoothly. If not, check for hair. Sometimes the motor goes bad or its lubrication dries out. If the spindle offers significant resistance, there's a mechanical problem of that nature.

The optical head provides two functions: tracking and data extraction. To get data, tracking has to be working. Either or both of these can be affected by oxidation or corrosion of the head's ribbon cable connectors. After a dead laser, this is the second most common head-related problem. Laptop drives, for some reason, are especially prone to this issue. Pull the cable at the board end and look at its metal fingers. If they're gold, wipe them with an alcohol-moistened swab. If they're silver, they probably have solder on them. These are the kind that cause the most trouble. With a magnifier, look for black pitting. Gently scrape it with the tip of an X-Acto knife and then use a swab to wipe away the tiny metal flakes. Clean with alcohol and reinsert. If there's a connector at the head end of the cable, do the same thing there.

Be aware that laser diodes are easily destroyed by static charges. Be sure you and your tools have touched ground just before you begin working on the cable. Reinsert the cable and check to see if proper operation has been restored. I've saved countless laptop drives with this procedure.

The primary output from the head is a signal called the *eye pattern*. See Figure 14-2. This signal is the actual, raw data being read from the disc. All players have a test point at the first preamplifier stage, which you can find by following the head's ribbon cable back to the board. The preamp will be a chip of low to medium density, and you'll see test points very near its pins. Often they're not labeled, but you can find the eye pattern by scoping the points; no other signal will look anything like it.

The eye pattern should be around 1 volt peak-to-peak. If it's much less than that, either the player is not tracking well or the laser is dim. Some amplitude wobble is normal as the disc spins, especially when the head is near the outer edge, but it shouldn't be more than 15 percent or so. If the amplitude dips a lot, expect skipping or dropping out. Weak lasers can cause this, as can problems with the focus servo. In players that have focus gain trimpots, you can sometimes compensate for a less-than-optimum laser by upping the gain a tad. Most newer players don't have servo adjustments, but it's worth looking for a trimpot labeled "F GAIN" or "FOC GAIN," just in case.

If all looks well but the player skips, check for binding in the sled. If it's well greased and clean, with no hair in the gears, there might be a tracking issue. With power off,

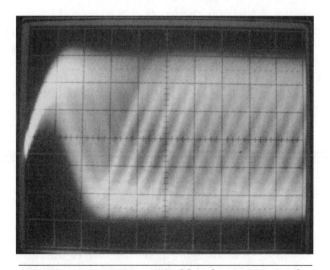

FIGURE 14-2   Laser optical head eye pattern of a CD at a sweep rate of 100 ns. These are fast signals! Use 50 ns for DVD eye patterns.

gently move the lens back and forth with a swab, checking that it moves freely. There could be dirt or hair there too.

Some players have tracking servo adjustments. Look for "TR GAIN" or "TRACKING GAIN" and try increasing it a little bit. The more hissing noise the lens makes, the higher the gain. You'll hear a "knee" above which the hiss will suddenly increase a great deal. Be sure to stay below that point or irregular tracking may occur.

If the machine is tracking the disc and the spindle is turning at the proper rate, there should be normal playback. Any other problems will be due to circuit failures that are probably more trouble to find than the machine is worth. Those kinds of problems are rare, though. Most players can be fixed with the procedures we've just examined.

Flat-Panel Displays

Unlike the old CRT technology, today's flat displays are matrixed. Each *pixel*, or picture element (a single dot), is addressed in X-Y fashion, so all of the circuitry required for scanning an electron beam over a screen is gone, and the associated service issues are different as well.

How They Work

In an LCD, a low voltage twists the molecules of a tiny bit of liquid-crystal material in each pixel, changing the orientation of its light polarization. In conjunction with a fixed polarizer at the front of the panel, that change of polarization darkens a pixel, making the pixel block the backlight's illumination. In a plasma display, a high voltage causes ionization of gas in each pixel, generating ultraviolet light that excites colored phosphors to generate visible light.

Today's displays have millions of pixels. A full HDTV display of 1920×1080 resolution contains 2,073,600 pixels, each with 3 subpixels of red, green and blue. That's 6,220,800 tiny dots, and every one has its own connection! How are all those addressed? There aren't millions of wires coming out the back, after all. The process works somewhat as it does with memory chips: row and column addresses are sent through decoders that fan out to the appropriate connections, eventually reaching each and every pixel through transparent, printed conductors at the edge of the glass. Using a grid formation allows far fewer connections than there are pixels; those 2 million pixels require just 6840 lines (1920 dots × 3 colors, plus the 1080 lines to select the rows). At the edges of the panel, the decoder chips make contact with the glass elements via ribbon cables affixed with pressure and conductive glue.

Virtually all LCDs made today are of the TFT, or thin-film transistor, variety. Instead of addressing the LCD elements directly, the exciting voltage pulses a transparent transistor behind each element. That enhancement lets the pixel store its state after it's been addressed, resulting in much higher contrast than if it only got pulsed and then left alone until the next frame of video came along.

Printing millions of functional transistors over the area of an entire screen requires very high-precision manufacturing. In the early years, TFT LCDs suffered from bad pixels; most had a few, and it was considered normal. Today's displays rarely ever show any stuck pixels. Nearly all panels, LCD or plasma, are 100-percent functional. It's pretty amazing, really.

What Can Go Wrong

Loss of a single connection out of thousands at the panel's edge results in an entire row or column that won't darken, leaving a bright line across or down the screen. Individual pixels can also fail, resulting in one dot of color that never moves on an LCD or a dark spot on a plasma. Plasma sets burn their phosphors when bright images don't move for an extended period, reducing the brightness of the affected pixels and leaving a ghost of the offending image. That happens most often with panels used for static display of airport schedules and such, but it also occurs at home with extended video game play and TV network "bugs," or logos, at the bottom of the screen.

The fluorescent backlights in LCDs are driven by inverters producing a fairly high voltage. They're just like the inverters in laptops, only bigger. Since the illumination provided always has to be as bright as the brightest picture could get, the inverters run pretty hard and hot, and are failure-prone. Some new sets use LED backlighting, eliminating the inverters. Those should last a good long time.

The high-voltage–generating circuitry in a plasma set also works very hard. Plasmas run rather hot, too, so they're more likely to experience thermal breakdown.

The power supplies providing the low-voltage, high-current power to run all this also generate some significant heat, leading to shortened component life and failure.

Is It Worth It?

If the panel develops a bad row or column, forget it. There is no way to repair that. I fixed a few early LCDs by making a plastic clamp to squeeze the ribbon cable's conductors against the glass, but in today's higher-density panels they're inaccessible, and the size scale would likely make such a crude repair impossible anyway.

If the glass is damaged, you can't repair it. Plenty of today's TVs get hit by a kid's toy or the family dog, rendering the sets useless. The cost of a new panel is usually more than the price of the TV.

Electronic problems like bad power supplies, blown inverters and failing capacitors can be successfully navigated. Luckily, those account for most of what you'll see. Computer monitors, especially, spend thousands of hours turned on, with the expected degradation of capacitors.

The Dangers Within

Plenty! In LCDs, watch out for the inverter and its output cables. They may have more than 1 KV on them. Plasmas are full of high voltage too, and it's fed to the panel's pixels, not just to a backlight lamp or two.

The actual liquid crystal material in an LCD is toxic and should not be handled. You'd come in contact with it only if the glass were broken.

How to Fix One

In plasma sets, look for bulging electrolytic power supply caps and bad connections. A total failure might indicate a blown chopper or other typical switching power supply issue. Beyond those, much of the set is made of specialized, high-voltage parts best left alone, unless the trouble is in a small-signal area like the tuner.

LCDs also have the usual power supply and cap problems, but their most frequent cause of failure is the backlight inverter. If you're not sure what an inverter looks like, see Figure 10-5 in Chapter 10. LCDs of any significant size have at least two lamps driven by multiple inverter circuits, often combined onto one board with output transformers at the ends.

If you turn on the set and it lights up for a second before the screen goes dark, one of the inverters has died. You get that one moment of light because the other section is running its own lamp until the micro senses the loss of the blown one and shuts them both off a second later.

The poor output transistors in an inverter work like dogs for countless hours, and eventually one of them shorts, taking the fuse on the inverter board with it. In most designs, each side will have its own fuse soldered to the board near the connector from the power supply or the microprocessor board. That's very helpful, because a blown fuse tells you which side of the board has quit. Try replacing the fuse first, even if just with a temporary arrangement employing two soldered wires, clip leads and a physically larger fuse of the same rating as the original. If you can't determine the original's rating, 3 amps is a reasonable value to try. Now and then you may find that the fuse has fatigued and failed but the rest of the circuitry is fine. If the lights come on and stay on, a new fuse is all you need. Be sure both lamps are working. The screen should be at normal brightness and evenly lit. If one end is significantly brighter than the other, one lamp is still out, and it's quite possible that the transistors on the blown side of the inverter became open from the momentary surge current when they shorted. So, they won't blow a new fuse, but they won't work either.

Take a peek at the inverter's transformer. If you see a burned spot anywhere, the transformer's insulation is damaged and the coil has arced over, either from one winding to the next or from a winding to the core. The increased current draw from arcing usually pops an output transistor. Changing the transistor does you no good, since the transformer will just kill the new one. If you see no burns, the transformer could still have internal shorts or arcing, but it's less likely.

If the transformer looks okay, you can change the inverter's transistors if you can find some. Often, they're oddball output components that aren't easy to locate. Sometimes you can substitute similar transistors, but they have to be a pretty close match, especially in their gain characteristics. Not enough gain will cause the transistors to run more in their linear region than fully saturated, and they'll get very hot and fail in a hurry. Too much gain can cause them to "ring," with the tops and bottoms of what should be a square wave having sine wave-like variations. That also puts them in their linear region and overheats them. If you do sub the transistors, scope their collectors or drains and compare what you see to the waveforms on the good side of the inverter. If they look a lot different, those transistors are not a suitable match. As long as the waveforms look like they're turning on and off all the way, and the inverter seems to work, let it run for a few minutes and then turn it off and touch the transistors. They might be warm, but shouldn't be too hot to touch. Compare them to the good side.

Hard Drives

Hard drives are in everything these days, from computers to MP3 players. Most people consider the drives irreparable—they either work or they don't—but that's not always true.

How They Work

Hard drives use a rotating platter coated with ferric material onto which is recorded digital data in the form of tiny regions of magnetism. The head flies on a cushion of air just a few wavelengths of light from the surface, suspended on an arm that flits at high speed around the disc, controlled by the drive's microprocessor as it reads and writes sectors of data. Unlike the heads in a video recorder, a hard drive head never touches the recording medium. The disc speed is held constant by a servo that monitors motor speed and locks it to a crystal reference.

What Can Go Wrong

The primary cause of hard drive failure is a head crash. The drives are well sealed, but even one very tiny piece of foreign material can stick to the disc surface, causing a disastrous scratch when the head hits it.

Wear or lesser scratching of the disc surface can result in bad sectors, areas of the disc that won't reliably return the data written to them. Most drives have a few, and they are hidden by a lookup table in the drive's ROM that avoids the known bad ones. If enough bad sectors accumulate, the drive is failing and will have a hard time processing data.

Stiction, the adhesive attraction caused by molecular forces between very highly polished surfaces, can make the head stick to the disc, especially in older, well-worn drives. This used to be much more of a problem than it is with modern drives. The symptom is that the drive has a hard time starting up. If it won't spin on its own but will start after a good slap, the head is probably sticking.

Another cause of hard starting is moisture inside. Because the flying of the head over the disc requires air pressure equalized to the outside world, hard drives have a small "breather" hole. The air coming in is carefully filtered, and a desiccant inside the drive absorbs moisture. You can hear its crystals moving around on some drives when you shake them. Eventually, the desiccant saturates and moisture builds up on the disc, crashing the head or making it stick.

The motor or its drive circuits can fail, in which case the disc won't turn no matter what.

The signals from the heads are very tiny, and a poor connection between the body of the drive and the heads can cause them to drop out or get too weak to read. The drive will recalibrate with a clacking noise, desperately trying to find the data. It may also write incorrectly, severely corrupting itself.

Is It Worth It?

Is there really anything you can do to repair a hard drive? Opening the case is out of the question; once you allow room air in, the drive is wrecked. Believe it or not, there is one thing you can try, and I've saved numerous drives with the procedure. It's fast and easy, and it doesn't involve breaking the seal.

The Dangers Within

No danger here. Voltages are low, and all the moving parts are sealed.

How to Fix One

If the drive is recalibrating often or returning errors, take off its circuit board and look at the connection points interfacing the board with the body of the unit. You'll see two sets—one for the motor and one for the heads. Rarely are the motor connections problematic. The signals from the heads, though, are so small that it doesn't take much resistance to lose them. See Figure 14-3.

Some brands of drives use lovely, gold-clad connectors that almost never cause trouble. Many manufacturers, though, save a few cents by replacing the connectors

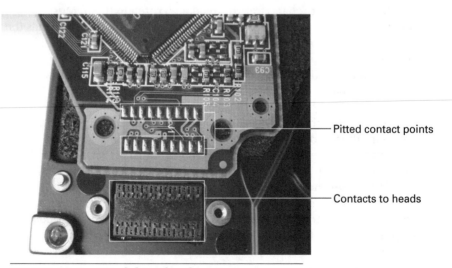

Pitted contact points

Contacts to heads

FIGURE 14-3 Hard drive head contacts

with sharp pins on the body of the drive that press into solder pads on the board. After a few years, the solder gets oxidized and its resistance goes up, impeding the delicate signals generated by the heads when they read data. If yours has the pads, look at the indentations made by the pins. Are they blackened at the center? That's the sign of this malady. Gently scrape off the oxidation with the tip of an X-Acto knife, and then wipe the pads with a dry swab to remove the metal flakes. Reassemble the drive and be prepared to be surprised! The darned thing just might work. If the drive has corrupted itself too much, it may be unrecoverable even though the electronics are now functional. Sometimes you can restore it only by reformatting, wiping out the data. Other times nothing works. But, heck, it wasn't doing you any good before anyway, right?

Laptop Computers

Laptop computers have largely replaced desktop machines in many people's homes and workplaces. They offer numerous advantages in terms of required space, power consumption and heat generation, but they're a lot more fragile.

How They Work

Laptops are functionally just like desktop computers, except that everything is much smaller and runs on less power, and the LCD monitor is built into the unit. Also, laptops incorporate power management systems for efficient charging and use of batteries, and their AC power supplies are not internal, except on some really old models.

What Can Go Wrong

Laptops are enormously complex, with most of the circuitry on the motherboard. That board has many layers and lots of LSI chips, along with a zillion tiny support components crammed together.

Some of those huge chips are connected to the board with a *ball grid array*, or BGA, which is a bunch of tiny, ball-shaped contacts soldered to pads underneath the chips. If you see a chip with no leads, it has a BGA. See Figure 14-4. BGAs provide hundreds of contacts in a small space, so they're used for microprocessors, video graphics chips and other very high-density devices. Some of those items run pretty hot, unfortunately, and can degrade their solder joints with flexure of the board and time, resulting in intermittent connections and a machine that keeps crashing.

How can you repair those things when you can't even see the contacts? It takes specialized rework equipment costing in the range of $50,000 to remove and resolder a BGA. There's just no way to do it at home. I foolishly tried resoldering one barely visible pin at the edge of a video graphics chip's BGA once, and succeeded only in destroying the chip and the motherboard. *Fuggedaboudit.*

FIGURE 14-4 LSI chip with BGA contacts underneath

Many motherboards sport a surprising number of surface-mount electrolytics, with the usual problems those cause. When changing them, be extra careful with the heat, since you're dealing with a multilayer board you could easily wreck. Review Chapter 12 for info on handling situations like that.

Battery charging is controlled with MOSFET power transistors. An open one will result in no charge reaching the battery. All modern laptops use *smart batteries* with their own microprocessors that tell the machine the state of charge, how many cycles have been used over the life of the pack, the battery's model number, and so on. If that micro gets scrambled, or the contacts between it and the laptop malfunction, the battery may not charge or even be recognized. Of course, worn-out cells will cause the same problems.

The power supply input jack is a frequent source of laptop misery. Pulling on the plug, tripping over the cord, and even just normal insertion and removal can crack the jack's solder joints, resulting in failure to charge the battery or no AC operation at all. If you have to wiggle the plug or push it to one side to make it work, the joints are cracked.

A dead backlight inverter is one of the most common laptop problems of all. The latest-generation machines use LED backlights, so they have no inverters, but most laptops out there still have fluorescent lamps. The lamp itself can get weak, and dropping the machine may break the bulb with no outward physical sign of damage. If the laptop has been dropped and suddenly won't light up, but you can see the image by shining a bright light on the screen, the long, thin lamp tube is probably shattered inside the bottom of the LCD.

The screen is connected to the body of the machine via cables running through the hinges. After the lid has been opened and closed hundreds of times over a few years, a cable can break. In most models, the inverter is located in the screen assembly to avoid having its high-voltage output wires going through the hinges. The power and control lines feeding it do go through them, and those are often the ones that break.

A bad RAM (memory) module can cause random crashes. Sometimes the module isn't actually defective, but its contacts have gotten oxidized, resulting in weak signal transfer to the motherboard. Bad RAM soldered on the motherboard will cause crashes too.

Usually, the hard drive and the optical drive coexist on the same system bus. If one of them malfunctions and hogs the bus when it's not being addressed, the machine will hang.

The keyboard can develop bad keys or entire bad rows. It can also drive the machine crazy with a stuck key that sends the same signal indefinitely.

Is It Worth It?

A bad motherboard is usually not worth your time, but there are exceptions. If you see signs of dying electrolytics, you can change those and probably restore normal operation. If the battery won't charge but everything else works, and you're certain the battery is good, you may be able to find the bad MOSFET and replace it. Those are big enough to remove and resolder. If there's some obscure logic failure or a bad BGA connection, it's toast.

Power supply jack problems are easy fixes, once you get to the darned thing! You'll spend far more time on disassembly than on the repair. In some laptops, removing the back gets you right where you need to be, but others require total disassembly to get anywhere near those precious solder joints.

Repair of a backlight inverter is tough, mostly because the tiny output transistors they use are hard to find, as are the soldered-in fuses. Replacing the inverter, however, is pretty easy on most models. Usually, all you need to do is open up the screen assembly, and there she is! Unplug the connectors, replace the board, and you're done—that is, unless the problem is a broken screen cable.

Ah, the screen cables. Very often the real reason the LCD won't light up is because the inverter's cable is broken. The problem can also be caused by a failure at the video graphics chip's BGA. So, it can be tough to tell whether the inverter, the cable or the motherboard is the true culprit. If moving the screen through its range of angles makes it turn on and off, one of the cable's wires is broken inside, and the ends are touching each other just enough to make contact when the cable is at certain positions.

Bad RAM can be easy or impossible, depending on whether it's a module or soldered to the motherboard. If it's soldered on, repair is unlikely.

Hard drives and optical drives are easily changed in some machines and hard to reach in others. The drives fail often enough that many manufacturers make them readily accessible, but some disregard that reality and bury the darned things so deep that it takes an hour or two to get to them.

The Dangers Within

Most of a laptop operates at low voltages, so it's pretty safe. The backlight inverter puts out a high voltage, so stay away from its output area. Switching converters on the motherboard can generate some other voltages that it's best to avoid, but they shouldn't be high enough to injure you. You could possibly get a shock if you touched the wrong point while the machine was running.

How to Fix One

Let's look at a few common laptop problems and how to approach their repair.

Crashes

If the machine works but crashes randomly, pull the RAM modules and use a dry swab to clean their contacts on both sides. Even if they're gold-clad, which most are, the swab may show a surprising amount of grayish dirt when you're done. Pop the modules back in and test.

If that doesn't solve the problem, use diagnostic software to check for a bad RAM module. If one comes up as bad, replace it. That's about all you can do. RAM is pretty reliable, but it does fail now and then. Some motherboards have RAM soldered on, with modules used only for expanding the memory above the stock configuration. If the motherboard RAM is bad, you're pretty much stuck. If it's an especially expensive laptop, you might be inclined to try replacing a RAM chip, but the size scale makes soldering difficult. Some of those chips aren't so tiny that it's impossible, though.

If the machine crashes so often that you can't even run the software, check for bulging or leaking capacitors on the motherboard.

Some laptops crash after they warm up. Usually, that means a bad chip or an intermittent solder joint, probably in a BGA somewhere under the microprocessor or the video graphics chip. I've seen hard drive controller chips do that too; the machine works fine until that chip warms up enough to malfunction, and then it won't read the drive or it corrupts the data.

Charging Problems

If the battery won't charge, check first to see if the battery is any good. The cells' chemistry wears out eventually. Also, the internal charge control circuitry can fail. Sometimes a battery stored so long that all of its charge has leaked away may refuse to start up because its internal micro won't signal its existence to the laptop without at least a little power to run. Some systems offer a battery resetting utility that can hunt for a dead battery and try some charge to see if it's there and working.

Short of cracking the battery open and applying a little charge directly to the cells, there's nothing you can do. And, frankly, I don't recommend doing that. It's rather difficult to accomplish without wrecking the battery, and the lithium-ion cells inside are pretty dangerous if a screwdriver pierces them while you're breaking open the plastic casing. Any breach of their seals can cause a fire hazard because lithium reacts violently with water, including water vapor in the air. Unfortunately, applying power to the pack's outside terminals won't work; the charge controller inside allows connection to the cells only when the laptop gives it the go-ahead.

If you're sure the battery is good, check to be sure the power supply is working! Many times the adapter fails, and it's just assumed that the computer is the culprit, when in fact it's not getting any power. If the adapter has an LED, it should be lit. Those LEDs are pretty much always driven by the adapter's output. So, if the light comes on at normal brightness, the supply is working. If it doesn't light up or is very dim, unplug the adapter from the computer and see if it comes back to normal. If so, there's a short in the computer dragging down the voltage. If not, either the supply is dead or its output cable has a short. I've seen that happen a few times, and repairing the cable restored normal operation, thanks to protection circuitry in the supply that prevented its destruction or a blown fuse from overcurrent.

Once you're sure the supply works, check for a jack issue on the laptop and in the supply's plug. Does the jack move when you wiggle the plug? Does indication of charging come and go when you do that? If you're watching for software indication of charging, remember that it may take 10 or 20 seconds for the operating system to recognize the change. Lights on the battery itself, if there are any, are a far better indication of whether it's receiving voltage. Even those may take a few seconds to respond, though, so wiggle slowly.

To ascertain whether the trouble is in the plug or the jack, hold the plug steady with one hand while moving the wire with the other. If the connection cuts in and out, the plug's the problem. If not, it's the jack. You may have to do this a few times to be sure, because it's easy to move the plug slightly in the jack while trying to hold it still.

When you're sure the power supply, jack and battery are okay, it's time to consider a motherboard problem. Power management on laptops is complicated, involving firmware (software encoded onto the machine's chips), system software and the power management unit, a specialized microprocessor used only for controlling the flow of power to the various parts of the machine. A problem with any one of these could prevent charging. If there's a reset procedure for the power management unit, try that.

If nothing works, open the machine and look in the area of the battery connector. Because significant current of up to several amps gets passed in charging, the power transistors controlling it are usually located near the connector to avoid having to waste space with wide circuit board traces that can handle the juice. The transistors could be on either side of the board. They'll be bigger than most of the components around them. Use a shield or other obvious ground point, like a chassis screw obviously connected to a wide trace on the board, for circuit ground. With the AC adapter connected, try scoping the transistors' terminals while you insert and remove the battery. If you find one with constant voltage on one terminal and a signal that changes state on the other when you pop the battery in, but nothing shows on the third terminal, the transistor is probably open.

Display Problems

Unlike a desktop computer monitor's DVI (Digital Video Interactive) or VGA (Video Graphics Array) connection, the interface from a laptop's LCD to the motherboard is specific to the particular make and model. Signals may be carried on a bundle of wires or a printed-circuit ribbon cable. Ribbons rarely break from normal flexure, although some get brittle after a number of years. Who keeps a laptop that long anyway? Wire bundles do break, resulting in all kinds of display symptoms.

Far and away, loss of backlight is the most common display failure. If you can boot up the machine normally and see the image on the screen by shining a bright light on it, the backlight has died. The three main causes are motherboard failure, cable breakage and a blown inverter. The fluorescent lamp may have broken if the machine took a fall, but that's less common.

This assumes there *is* a lamp. Even in newer laptops using LED backlighting, though, operation still depends on functioning cables from motherboard to screen.

With the laptop running, gently move the screen back and forth through its entire range of angles. If you see even a flicker of backlight, the problem is almost certainly in the inverter's cable. Most inverters have at least five wires going to them, and sometimes more. Usually, there's +5 V, +12 V, on/off control, brightness and ground. Brightness is set by varying the duty cycle of a fast square wave. The higher the brightness, the longer the waveform stays up during each cycle. That arrangement allows the motherboard to control the brightness in purely digital fashion, with no varying analog voltages. If the wire breaks, though, the inverter will go dark, even though all other inputs are working. In fact, some models omit the on/off control wire and just turn off the pulse train to shut down the inverter when you close the screen or set the brightness to minimum. Loss of any of the other wires will kill the backlight too, of course.

Inverters work hard; most the warmth you feel along the bottom edge of a laptop's screen comes from the inverter's output stage. So, as you might expect, the transistors blow. That's especially likely when the screen has been run at full brightness for a few years. The heating and cooling cycle eventually kills those transistors.

Unlike desktop displays, laptops usually have only one lamp and one inverter. Open up the screen assembly and you should see it. It'll look just like the ones in desktop LCDs and TVs, only smaller. It may be covered in tape or with a shield. Very close to the inverter's input connector (not the two wires going into the screen—that's the high-voltage output for the lamp), you'll find a fuse soldered to the board. With power disconnected, check it with your DMM. If the fuse is open, either the inverter has blown or the lamp is broken inside the LCD. Unless the machine got dropped, assume the inverter's output transistors shorted and blew the fuse. As described earlier, you can try replacing those, but they're hard to get. The easiest route is to replace the entire inverter board. Some manufacturers won't even sell you an inverter; they want you to replace the entire screen assembly, which probably costs more than the laptop is worth. Check eBay and online parts houses for good deals. The first time I ran into this problem, I opted against the manufacturer's $700 screen assembly replacement and found a new inverter online for $40. It worked just fine.

If the fuse is still good, a problem with the inverter is less likely. One of those pesky transistors could have opened and not blown the fuse, but they usually short. The cable may be broken, even if the swivel test didn't turn anything up, or there could be a serious motherboard problem like a bad graphics chip or a broken BGA connection. Try pressing on the graphics chip. If the light flickers on, the machine crashes, or any other changes occur, there's your answer: its BGA is intermittent.

Testing the cable isn't hard. Unplug it at both ends and use your DMM to check for continuity. To get a connection to those tiny holes in the connectors, connect a clip lead to a small component from your stash and push its lead into the connector, with the other end going to the DMM. Needless to say, you want to insert the same end of the component to which your lead is clipped! Just let the other end of the part hang.

If the cable is good and the inverter's fuse is good, it's likely the motherboard isn't turning the screen on, and you're probably not going to be able to fix it without replacing the board. A lot of laptops get dropped or otherwise abused, and people sell them off for parts when the screen is cracked. The motherboard rarely gets damaged in a fall, so you may be able to scare up a parts unit online and swap out the boards. Just avoid buying any machine that's had liquid spilled into it.

If the screen lights up but the video isn't normal, plug an external monitor into the machine and see if video works properly on that. If not, the motherboard has a serious problem at the graphics chip. If it looks okay externally, either the video cable or the screen itself is causing the trouble.

A single bad line on the display cannot be the fault of the cable, because no one wire is specific to such a small area of the screen. A bad line or two is caused by the row and column drivers inside the LCD or their connections to the glass. You can't fix this, but finding a reasonably priced replacement screen online is easy enough for many laptop models. Changing it is entirely a mechanical job; no soldering will be required. Take apart the screen bezel, get the LCD in the frame, plug in its connector and you're done. You shouldn't even have to open the main body of the computer.

When video is severely distorted, with large areas of the screen a total mess, suspect the cable or the graphics chip. A bad screen can cause this too, but it's less likely unless you see obvious cracks from a fall. I once worked on a laptop with video that started shaking back and forth after it had been on for about half an hour. Eventually, the image would tear, looking a lot like an analog TV with its horizontal hold misadjusted. I proved the fault was with the graphics chip by spraying component cooler on its heatsink. As soon as the chip cooled even a little bit, video returned to normal for a few minutes, until it got hot again.

Before you give up, check the video cable the same way as with the inverter cable. A broken connection there will cause absolute havoc on the screen. Also, some LCDs have thin circuit boards on the back, with ribbon cables connecting them to the row and column driver chips along the screen's edges. If they use sockets, check the fingers on those cables to be sure they're not oxidized. Clean them and test the screen again. Beware the thin, printed conductors wrapping around the edges of the screen! Those are the connections to the drivers, and they're especially fragile because of their density. Pressing on one is likely to ruin the LCD. Keep that in mind when installing a new panel, too.

Drive Problems

As described in this chapter's section on hard drives, there is an issue with some brands of drives, especially after they've been in use for a few years. The contacts from the board to the head assembly inside the metal casing get oxidized, causing read failures and sometimes severe data corruption. For some reason, laptop drives seem especially susceptible to this problem. If yours is recalibrating a lot, having trouble reading and returning errors, it's worth taking it out and trying the procedure described in that section.

The 5 VDC power to the drive needs to be quite steady and clean for the drive to function properly. Drives pull up to 500 milliamps, which is not insignificant. If the motherboard's electrolytics are starting to get weak, the voltage may dip or develop spikes when the drive turns on, making it corrupt itself or causing malfunctions in other areas of the board. As the drive spins up from a dead stop, current draw can be a full amp for a second or so. See if malfunctions occur at the moment spinup begins. If so, suspect voltage regulation or bad caps on the motherboard.

The optical drive eats around the same amount of current and can cause similar trouble. The information in the section on CD and DVD players, also in this chapter, applies to computer optical drives as well. In particular, the problem with oxidized ribbon connections to the laser head occurs more often in these drives than in shelf-style DVD players. If the drive has trouble reading discs, try cleaning the lens and the cable contacts. Very often, that'll bring it back to health.

Other Problems

The Wi-Fi antenna cable on laptops with internal wireless cards goes through the hinges, with the antenna in the screen assembly. A broken coaxial cable will cause severely reduced wireless range. To check the cable, unplug it from the Wi-Fi card and use your DMM. Many machines have two antennas, with a small duplexing board in the screen assembly, so you can't test from the card end of the cable right to one of the antennas. Be sure to check from the card connector to the duplexer. If you follow the cables from the antennas, which are little flat things next to the screen, they'll meet at the duplexer. The third cable, going down to the hinges, is the one you want. Test from its solder contact on the duplexer back to the other end inside the main body of the laptop. Be sure to check both the shield and the center conductor, as either or both could be broken. Usually, it's the center conductor that breaks.

Laptop keyboards are mostly mechanical, with a conductive rubber button under each key. The connections are arranged in a grid, but the irregular layout can lead to some pretty obscure patterns. You can't assume that it's all rows and columns.

When one key stops working, it's a contact problem underneath. Time and oxidation can do it, but more frequently someone has spilled liquid into the keyboard. If you're really intrepid, you might be able to take apart the keyboard and clean it out, but it's a lot of trouble, and keyboards for most machines can be had pretty inexpensively. Replacing one is just a matter of popping it off, disconnecting the ribbon cable, and installing the new one.

If a whole bunch of keys dies, and there hasn't been a liquid spill, check the ribbon cable connections at the motherboard, because one entire line may be out. If not, there could be a motherboard problem with the keyboard decoder. Or, the break might be in the keyboard.

To check it, remove the keyboard and do a continuity test on every combination of lines coming from the keyboard, keeping in mind that good contacts may show as many as a few dozen ohms, as is normal for conductive rubber switches. There aren't that many lines, perhaps six or eight, so the test isn't that rough. If a bad key produces no continuity on any combination, the keyboard is the problem.

If the machine goes nuts, acting like someone is typing the same key over and over, that just might be the case! That "someone" was the person who spilled coffee or soda into the keyboard, shorting one or more of the contacts with conductive goo. Disconnect the keyboard and fire up the machine. If the stuck key goes away, you know somebody got sloppy with the drinks. This happens quite often, and the poor computer will act like it's mondo loco, when all it really needs is some peace and quiet from its keyboard connector.

MP3 Players

Being portable, MP3 players are subject to mechanical damage from being bounced around and dropped. The parts are small, making service challenging, but most players can be fixed.

How They Work

MP3 players retrieve blocks of data from a storage medium, either flash memory (the nonvolatile memory like that found in pen drives) or a mechanical hard drive. The data is quickly read into memory and then read back out at the rate required for conversion to uncompressed audio. When the memory buffer is nearly empty, another block of data is read from storage to refill it, providing continuous playback. The reconstituted audio data is fed to the digital-to-analog converter, or DAC, filtered to remove digital artifacts and boosted to headphone level with a small stereo amplifier.

MP4 video players work the same way, except that they have decompression chips for video data, along with an LCD screen capable of video display.

Overseeing all this is a microprocessor that selects tracks, minds the buffer, and extracts and displays information such as song title, artist and data rate.

What Can Go Wrong

Most MP3 player problems involve broken solder joints on their headphone and power jacks, failing batteries, cracked displays, bad hard drives and drive controller chips, and loose cables. People sometimes sit on the players, breaking the screen or crushing the case.

Is It Worth It?

Repairing broken joints on jacks costs nothing but your time to disassemble the unit and resolder them. Batteries for popular, more expensive players are available at low cost online. Broken screens can be found there too, but if the case is badly bent or the circuit board is damaged, it's not worth trying to repair the machine. The small hard drives used in MP3 players are a little more expensive than equivalent-capacity laptop or desktop drives, but they're not prohibitive. A bad drive controller chip is irreparable without replacing the board, which is probably not worth the cost.

The Dangers Within

No dangers, except possibly from shorting the battery wires. Lithium-ion batteries can supply a fair amount of current into a short. The batteries get hot when they do, too, and could burst if the short isn't resolved quickly.

How to Fix One

Opening the cases of some players is the hardest part. While some come apart the traditional way, with a few tiny screws, others are snapped together snugly and require a shim to separate the halves. Disassembly instructions for the more popular players are available online, as are plastic shim tools. Very often, the purveyors of batteries and drives for MP3 players include the tool with their products. Be sure to open the case gently, as ribbon cables may connect the halves.

Before doing any soldering, disconnect the battery. Most batteries plug in on a connector, so pull it. Some players have directly soldered battery wires. Carefully unsolder the negative (black) wire and put a piece of electrical tape over the bare end so it can't short against anything.

If the hard drive is clacking and not retrieving data, the drive itself or the controller chip may be at fault. The only way to tell is to swap out the drive. The controllers use the same BGA connections that so many laptop chips have, with the same problem of broken solder joints under the components. There's no way to fix them, but sometimes you can insert a shim of some sort to press against the chip and keep the connection working for awhile. That sort of "repair" doesn't last long, though.

Screen and drive replacement are straightforward. Be sure ribbon connector latches are securely closed; they tend to loosen in items that get bounced around. I like to put a little nail polish on the edges, at the latch tabs, just to be sure they stay put. If you do that, brush on only a small amount. You don't want it dripping into the connector. Let it dry before closing the case, so the outgassing won't remain inside.

VCRs and Camcorders

Tape-based video recorders are complex machines with many interactions between their various sections. Servicing them can be fascinating and challenging.

How They Work

VCRs and tape-based camcorders, both analog and digital, record very high-speed signals using the *helical scan* method, in which the tape is wrapped around a rotating drum, also called a cylinder. Protruding slightly from the drum are two video heads that scan the tape in a diagonal pattern, laying down very thin tracks next to each other, efficiently utilizing the width of the tape as well as its length to pack in lots of information. Helical-scan recorders are some of the most mechanically complex consumer devices ever sold. As a dying technology, they're cheap now, but their high cost in the early years of home video recording was not from price gouging. The mechanical precision required in their manufacture, especially of the head drum and heads, was in the microns. Those babies were hard to make! It took 20 years of research and development to bring VCRs to the masses, and their eventual low cost was due only to the economy of scale.

Analog recorders are more electronically complicated than their digital counterparts. First, the amplitude variations and noise inherent in weak, fast signals coming from tape necessitate the use of FM recording. The video signal is used to frequency modulate an oscillator, and the resulting RF signal is then recorded. On playback, FM demodulation extracts information only from the frequency variation of the carrier signal, ignoring noise and wobbling amplitude. As with FM radio, the technique permits an inherently noisy channel to provide a clean signal.

TV signals are extremely time-sensitive, and no mechanical system can recover them from a recorded tape without timing errors rendering them from wobbly to useless. So, correction schemes involving phase-locked loops (PLLs) and, much later, digital timebase stabilization were developed. Color was especially difficult to reproduce, as even nanosecond-level jitters would wipe it out. A double PLL system—one for gross errors and one for fine adjustment—finally solved that problem.

Tape isn't perfect; it has microscopic areas where the magnetic oxide flakes off or is damaged by dust. On playback, the signal randomly drops out for very short periods as the heads lose contact with the tape. In an audio recorder, such dropouts are too small to be heard, and they also don't extend across the width of the entire track, so they have no effect beyond slightly increasing the noise level. With the dense, narrow tracks of helical recording, dropouts trash entire lines of video. The *dropout compensator*, an analog delay circuit storing a line of video, fills in missing lines with the previous one, rendering most dropouts invisible.

Digital helical machines only need to get bits on and off the tape, and they tolerate and correct timing errors by reading the data into memory and then clocking it back out steadily before converting it to an analog output signal. So, the timing correction systems in analog machines aren't needed, and a simpler PLL system suffices. The data rate is quite high, however, and dropouts are still a problem. Digital error correction codes are used to calculate missing bits and fill in for what gets lost. The MiniDV format uses image compression similar to that in a JPEG, converting each frame to its compressed form and recording it separately. Newer formats using MPEG-2 or MPEG-4 compression periodically record *key frames*, which are complete frames, and then store only the changes between frames until it's time for another key frame. This more advanced technique results in far fewer required bits for a given picture quality level, but it's less suitable to tape recording because lost bits can affect many frames, not just the one in which they occur. MPEG recording is used in virtually all card-style and hard drive recorders, though, because recovery of all the bits is pretty well assured, and its higher storage efficiency is crucial. A few MPEG tape camcorders have been marketed, but they've never caught on.

In both analog and digital recorders, servos are used to control tape and head drum motion. In analog recording, the heads are time-aligned with the video frames by a *head drum servo* system to avoid having the machine switch from one head to the other during the frame. If you've ever rolled the vertical hold on a TV while playing a tape, you've seen a line of distortion just above the vertical sync bar. That's the *switching line*, where the machine switched between heads while recording. Imagine that in the middle of the picture! Without a servo to keep it near the sync and off the screen, the ugly line would wander through the frame.

To keep the heads centered on the tracks during playback, either head drum rotation or the tape motion through the machine must be controlled. Most VHS machines lock the heads to a crystal to keep their rotation steady, and control the pulling of the tape with a *capstan servo*, positioning the tape tracks under the heads as they fly by. It can also be done the other way, keeping the tape speed steady and adjusting head rotation, and some recorders use that method.

In VHS, the video track position on the tape is sensed using a *control track* of pulses recorded along the edge of the tape, one pulse for each head drum rotation. Newer formats like 8mm, Hi-8mm and MiniDV use signals recorded into the helical tracks themselves, and have no separate control track. While early VHS recorders sported manual tracking controls, modern units discern the correct setting automatically, taking a few seconds to find the best alignment. Later formats without control tracks lock up much faster. However it's done, the servo must center the rotating heads over those tracks for playback to occur, and servo problems are frequent causes of helical recorder failures.

The mechanical sections of VCRs are much like those of any tape recorder, except for the necessity of pulling the tape out of the cassette and wrapping it halfway around the head drum. This looks easy, but doing it reliably without damaging the tape was one of the greatest obstacles in the development of video cassette machines. Loading problems are also frequent repair issues, especially in camcorders, with their tiny mechanisms that get tossed around in normal use.

The rest of a VCR is a TV receiver, with a tuner, audio and video sections. Early machines recorded audio along the tape's upper edge in linear fashion, like any analog audio recorder. The hi-fi system added frequency-modulated carriers for audio, recording them with rotating heads on the drum, along with the video tracks. All VHS hi-fi machines still record a monaural, linear audio track for compatibility with non-hi-fi units, but later formats never included one.

The rest of a camcorder is a video camera, typically with a motorized zoom lens, autofocus and all kinds of signal processing. Modern camcorders, both analog and digital, use digital signal processing in their camera sections, eliminating all the trimpots and wads of circuitry used in older analog designs.

What Can Go Wrong

Most VCR and camcorder problems result from mechanical issues with the loading mechanisms and tape drives. Clogged video heads are common, and cleaning them requires care and a gentle, steady hand.

The tape has to be taken in, properly seated and threaded around the head drum. Broken nylon gears are typical in VHS machines. Bent cassette carriages and loading arms cause a lot of camcorder failures because the parts are small, thin and easily deformed.

The tape path gets dirty from tape oxide residue and periodically must be cleaned. The tension arm with its felt band around the supply spindle, from which the tape is fed, often gets bent just enough to misalign the tape path around the heads. That path is critical and tough to realign.

Video heads wear out, making proper tracking increasingly difficult and unreliable. In analog recorders, the drum spins at 1800 RPM. In a MiniDV machine, it zips along at 9000 RPM. So, for each second of operation, those heads are seeing a lot of tape! It's a wonder they last as long as they do. It takes years of frequent use to wear out VHS heads and at least a couple to grind down MiniDV heads.

The entire machine's operations are directed by a microprocessor with many inputs. Various parts of the loading mechanism report back to the micro via leaf switch, mode switch or optical sensor so it can move them in the proper sequence. The micro also keeps tabs on whether the spindles are moving and if the head drum and capstan motor are rotating at proper speed. A fault in any of these areas can trigger a shutdown. Those protections are a legacy of the early videotape years, when recorders, video heads and even the cassettes were quite expensive.

Is It Worth It?

VCRs are obsolete and cheap, but you might try to save one because you have tapes you want to continue to be able to view. Perhaps you even have a Betamax and some priceless home movies from long ago you need to dub to a DVD recorder. Camcorders range from throwaways to very expensive pro or semipro machines, so some of them may be more worth the effort to repair.

If the heads or other mechanical parts are worn out or broken, forget it, at least with VCRs and low-end camcorders. As with CD and DVD players, sources for video recorder parts are gone because they'd cost more than the machines they fit, if you consider the price of labor to change them. Service shops can still get camcorder parts, but it's unlikely you'll have access to them.

Heads are easy enough to clean, and tape path realignment of analog recorders, while something of an art form, can be achieved with a known good tape, a steady hand and an oscilloscope. Alignment of a pocket-sized digital camcorder's path is pretty nasty, because it's unlikely you can get to the required signal's test point without specialized factory equipment. I've had some luck doing it by trial and error, carefully noting the extremes at which picture blocks appeared and then setting the tape guides to the middle of the range.

The Dangers Within

The rotating head drum is easily damaged, and it can cut you if you contact its edge while it's whizzing around at full speed. Beyond that and keeping your fingers out of the way of the loading gears and such, there isn't much to worry about besides the usual power supply stuff. If you're not working on the supply, cover it to avoid inadvertently touching hazardous voltages.

How to Fix One

Let's look at some of the most common problems and how to solve them.

Stuck Tapes

If a tape is stuck in the machine, don't tear it out or you'll do more damage than you were trying to fix in the first place. Stuck tapes mean either a power supply problem or a mechanical failure. Most of the time, the rubber wheel driving the spindles has dried or worn out. Sometimes that results in a tape spill, and loops of tape can get wound around the guides and loading arm. If you rip out the tape, you'll probably bend or break those parts, and that's the end of the recorder.

Many machines use a belt to drive the wormgears that raise and lower the cassette carriage. Some also use one to move the loading arms that pull the tape from the cassette and wrap it around the head drum. When the belt stretches with age, the mechanism gets stuck and the micro goes into protection mode, shutting down the machine. VCR belts and wheels, once readily available, are fast disappearing. You may be able to retrofit a part from another machine. It's tempting to try a rubber band, but those stretch too much to be useful. The loading functions aren't so exacting, though, that a belt close in size but not a perfect fit might not work.

Take a look at Figure 14-5. Before trying to remove a tape, check the position of the loading arms. Are they retracted, with the tape not protruding from the cassette shell? That's the best-case scenario. If you're lucky enough to find that, hunt around the sides of the cassette carriage for the loading motor and wormgear. Look for a belt from the motor to the first gear. With all power disconnected, try turning the gear by hand. If it won't turn one way, try the other. In one direction, it should begin to eject the tape. Keep turning it until the cassette can be removed normally.

If the loading mechanism is engaged, with the arms out in the threaded position or any position past fully retracted, you have a bigger problem. Sometimes the loading arms are properly retracted but the tape is out of the cassette anyway, because the spindles didn't turn to take it up while it was being unthreaded.

Either way, the problem is the same. While you might be able to remove the cassette with the procedure I just described, it's unlikely you'll get it out without destroying the tape. That's no biggie if it's just a recording of some old TV show, but if it's your only record of little Jenny's first birthday party, you don't want to lose it! Once videotape is creased or stretched, there is no way to flatten it back out and get a picture off of it. And if the damage is severe, running the bad spot through a machine

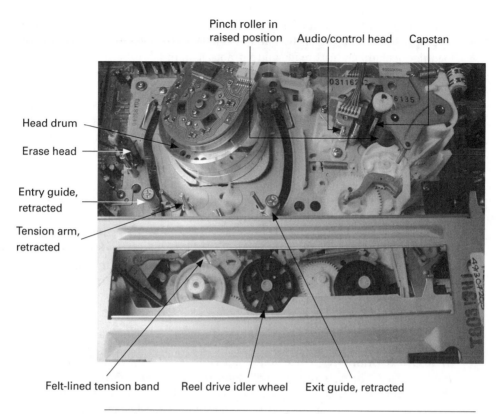

FIGURE 14-5   VHS transport mechanism

later on may clog the heads every time you try it, making it hard to play the rest of the recording.

If you don't care about the tape, just cut it away and pull the shreds out of the mechanism. Then use the manual gear-turning procedure to eject the cassette. If the tape does matter, it's worth trying to remove it intact.

If things got stuck with the tape threaded, it may already be damaged, but at least you can try to avoid causing further carnage. Skin oils ruin videotape, so don't touch it with your bare hands. Put on a pair of fresh disposable rubber gloves, being sure to use a type that is *not* covered in talcum powder or lubricant of any sort. Try to extricate the tape from the mechanism as gently as you can. If there's grease along the bottom of moving parts, where they contact the chassis, be especially careful not to let the tape touch it. Playing a greasy tape can wreck a video recorder, and cleaning the grease from the tape is pretty much impossible without causing serious damage to its recorded contents.

During play, the tape is held against the rotating capstan shaft by a rubber roller called the pinch roller. Mechanisms that lower it from above make removal of tapes without damage especially difficult if the roller is in its lowered, ready-to-play position. To get a tape out of one of those requires manually turning the loading motor assembly's first driven gear by hand, just as with the cassette carriage motor, until the loading mechanism unthreads and retracts. The pinch roller will rise, freeing the tape. Look for the loading motor on the underside of the tape transport chassis. I've seen a few on top, but not many.

Once you've extricated the tape from the mechanism, you need to turn either of the spindles to take it up before you attempt to eject the tape. It's a good-old catch 22: you can't reach them with the cassette in place, and you can't remove the cassette until you turn them. Place the transport on its side. On the back you'll see a pulley positioned between where the two spindles are on the top side. Usually, it's driven by a long belt from the capstan motor, though some fancy models use a separate motor to turn it. Try to turn it by hand, watching the cassette's reels to see when one is turning the right way to take up the tape. The supply reel, on the left, should turn counterclockwise. If the takeup reel, on the right, turns instead, it should go clockwise. Turn the pulley while carefully guiding the tape with your other hand, keeping it away from grease and from getting caught on the mechanism. When it's fully wound into the cassette, eject it by turning the carriage motor's gear.

If the tape is damaged, which is likely, avoid playing that spot on another machine, lest you wind up with another repair job on your hands! It's best to start playback after the damaged area.

Why Did It Happen?   Once you've gotten the tape out, check for why it got stuck in the first place. The cassette itself will not cause this problem unless it has a label on it that peels off and jams the machine. Most of the time, stuck tapes are the recorder's fault.

Typically, the rubber wheel driving the spindles has lost friction. If that long belt on the underside has stretched or gotten dirty enough to slip, it'll cause the same problem. The telltale symptom is that the mechanism has properly retracted but the tape is still out of the cassette. Machines with bad belts or wheels often have trouble rewinding, too. Although replacement parts are hard to find these days, it may be possible to save the old one. Try cleaning it with a swap moistened with a small amount of naphtha. Don't saturate the swab! Just a drop or two will do it. Also clean the mating surfaces of both spindles. Let everything dry for a minute or so, and then turn the pulley on the other side of the transport while holding whichever spindle tries to turn. See if there's a good grip. Turn the pulley the other way and test the other spindle in the same fashion. If the repair is successful, don't expect it to last for years. Once the rubber wears out, there's no keeping it alive for a long time. It might see you through dubbing some tapes, though.

Other Loading Problems

The accepting, loading, unloading and ejecting of tapes are coordinated by the unit's microprocessor. Each process requires several steps, and the micro has to know where the mechanical parts are to perform the required sequence successfully. Buried somewhere in the mechanism, usually on the underside, is a *mode switch* with multiple contacts that connect when various parts of the mechanism reach their destinations. Most mode switches are rotary, with fingers that rub against a set of printed-circuit contacts. Some are slide switches. Look for a bunch of wires or a ribbon cable going to the switch.

Dirty or oxidized mode switch contacts will seriously confuse the machine, causing all manner of odd mechanical behavior, from random shutdowns to stuck loading arms and out-of-sequence movements that ruin tapes. A little contact cleaner spray into the switch often does wonders. Just be careful not to let it get on the rubber parts or it will lubricate them too much for proper traction, and you'll have to clean it off. After spraying, run the mechanism through its loading and unloading paces at least half a dozen times to help the spray clear the dirt.

Cleaning

Any time you service a tape recorder, audio or video, clean the tape path. A dirty path will cause symptoms ranging from snowy playback to none at all. Even if playback is okay, clean the machine anyway. Helical recorders are especially prone to problems from tape oxide and other dirt because the alignment of the tape with the head drum is critical down to absurdly small tolerances. As gunk builds up along the drum's track, the lower edge on which the tape rides, it forces the tape slightly upward, disturbing the alignment and causing mistracking. Also, the contact surface of the video heads themselves is very small, and it doesn't take much to come between it and the tape. Any loss of contact results in nearly complete loss of signal.

To clean the tape path, you'll need some swabs, isopropyl alcohol and a sheet of white printer paper. Dip a swab in alcohol and clean the loading guides that pull the tape out of the cassette and around the head drum. Pay extra attention to the ends against which the edges of the tape rub. Those get the dirtiest and are also the most alignment-critical. Then clean the tension arm and the erase head, on the left. Next, clean the stationary audio/control head on the right, the capstan and the pinch roller. Once you've cleaned the roller, throw away that swab.

That's the easy stuff. The head drum is the most delicate part of the machine, and cleaning it without doing damage to the heads, also called the *head tips*, requires extra care. Turn it slowly from above while you look at the slit where the rotating section meets the stationary base. As it turns, you'll see a small, rectangular-ish hole pass by. Sticking ever so slightly out of the hole is a video head. See Figure 14-6. The head is very narrow. It's made of a strong but brittle ferrite material that can endure thousands of hours of high-speed rubbing against the tape. It *cannot* withstand much up-and-down pressure at all! If you push up or down on it, you will snap it off, and that's the end of the recorder. You'll find anywhere from two to six holes with heads, depending on how fancy a model the recorder happens to be.

To clean the drum, first dip a fresh swab in alcohol and clean the edge of the track, which runs diagonally along the bottom, from its highest point at the left to its lowest at the right. This is where the tape rides, and it needs to be really clean for proper tracking to be assured. Then clean everything between the track and the slit, carefully avoiding the video heads. (Rotate them out of the way as you go.) Once that's done, dip another swab and clean the rotating drum everywhere *except* very near the video heads or the heads themselves. Avoid those babies! Clean the drum by holding the swab and turning the drum against it from above, being careful not to get finger oils on the outer surface, which contacts the tape.

FIGURE 14-6 Close-up of video head tip

Now comes the fun part. Take the printer paper and fold it in half. Wet an inch or two of one side with alcohol. If alcohol is dripping from it, let that drip off, and blot the paper with another piece. Turn the drum so no video heads face you. Press the wet area of the paper against the drum, but not hard. Make sure some of that pressure is at the slit. Now turn the drum slowly so the video heads will rub against the paper as they pass by. Rotate the drum several times, being careful not to put pressure on the heads in an up or down direction. Remove the paper and let the drum dry. You should now have some black streaks on the paper and one clean VCR!

Tracking Problems

If the tape doesn't track accurately, playback signal will be lost through part of the head's sweep across the tape, and you'll see snow somewhere in the picture, or the image will jitter vertically. There are several varieties of tracking maladies. If the machine won't track a tape it recorded, there's a loss of head contact from worn-out heads, dirt, misalignment of the entry and exit guides controlling the tape position on the drum or a lack of proper tape tension. Check the tape tension by gently moving the tension arm on the left. As you press it toward the left, the buzzing of the heads against the tape should increase significantly. As you push it right, the buzzing should get quiet. If tension is too low, there won't be much difference between its normal position and when you push it toward the right. The arm connects to a band wrapped around the left spindle. Between the spindle and the band is a coating of felt. The felt may be worn, or the mating surface on the spindle might be dirty. Clean the spindle's surface with alcohol and check again.

Sometimes the tension arm gets bent upward a little from pulling against the tape, so the tape doesn't sit properly on the guides. Gently bending it back may solve the problem. The tension arm should sit straight up and down. Bend it only if it is obviously tilted. Even then, go very carefully, because it's easy to wind up with serious alignment problems if you get it out of whack.

If tapes made on the machine look fine but it won't play those recorded on other VCRs, there are two possible reasons: one of the guides may be off, or the position of the audio/control head may be incorrect. To test, play a tape recorded at the slowest speed on a VCR whose alignment you trust. Run the tracking through its range using the buttons on the remote (unless it's a really old VCR, in which case it may have a knob on the front panel). As the tracking shifts, snow will appear. If it's mostly at the top or bottom of the picture, and it moves up or down a lot when you change the tracking, it's a guide problem. If it appears across the entire picture at once, or nearly so, the audio/control head's position is off.

Adjusting the audio/control (a/c) head is easy. Set the machine's tracking control to the center of its range. On one side of the a/c head, you'll see a cone-shaped screw. Turning it will move the head left or right without changing its tilt. Adjust the screw just a little. If the picture gets better, turn it some more, until the image looks good. If it gets worse, turn it the other way. Try not to turn it more than a few degrees either way.

Run the tracking through its range again and see if picture noise appears at approximately equal distance away from the center position. Adjust the audio/control head until that's the case. You should wind up with optimum tracking at or very near the center of the machine's tracking adjustment range.

Adjusting the guides is a much more involved affair. To set them, you need a tape recorded at the lowest speed on a machine you trust, as before. Put it in and set the tracking for the least snowy picture, even if one end has noise. Try to minimize the overall noise. It's better to have a strong picture with bad noise at one end than a slightly noisy image across the whole screen.

Warm up the ol' oscilloscope and set it for 200 mv/div and a sweep rate of 5 ms. Look for two cables coming out of the head drum, one from the top and one from the back. One of them will lead to a shielded area. It may be a separate metal box, or it may just be a shield over the circuit board. That's the video head preamp. On it or nearby will be some test points on a connector that has nothing inserted into it. Using the metal chassis as ground, scope the pins and you'll find a signal that looks like the upper waveform in Figure 14-7. This is the *RF envelope*, an amplified rendition of what's actually coming off the heads. Connect your probe to it, being careful not to short adjacent pins with the clip. Scope triggering will be unstable, so the waveform will jitter from side to side.

To get a stationary trace, set the scope's second channel to 5 volts/div and the trigger to channel 2. Enable chop mode and scope some of the other pins on that connector until you find a square wave. Adjust the trigger level control for a stable sweep. You should now be able to see a rock-solid envelope, with the triggering square wave underneath. The envelope contains the signals from both heads, one after the other, and you can view the interruption between them at the point the square wave changes states.

The start of the head sweep, and thus of the envelope waveform, is on the left. In the recorder, that corresponds to the entry guide just to the left of the drum. If the envelope is lower at the start of each sweep, that's the misaligned guide. Before you adjust it, take a good look at how the tape rides on it. If it's high, not touching the

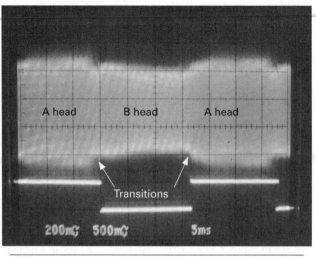

FIGURE 14-7 RF envelope

guide at the bottom, the tension arm may be bent upward a little bit, and you should gently bend it straight before trying to align the guide. Watch how the tape rides on the guide to get this correct. See Figure 14-8.

If the waveform is low on the other side, the problem is the exit guide, on the right. To align either guide, you need to unthread the tape and slightly open the setscrew at

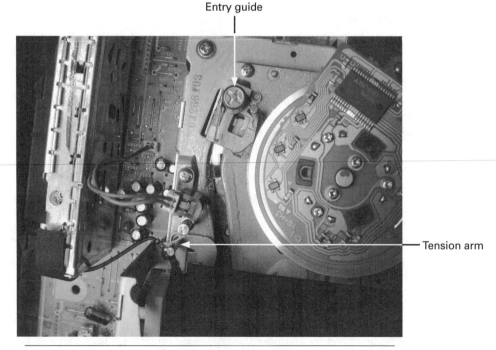

FIGURE 14-8 Entry side with tension arm

the bottom of the guide, if there is one. Don't make it loose; you need some friction there. If there's no setscrew, you can proceed without worrying about it. Hit play. When the tape is loaded and playback begins, turn the split screw adjustment at the top of the guide to lower or raise it. Tiny amounts suffice here; it's unlikely you'll need to turn it more than a few degrees.

The correct adjustment tool is a special split screwdriver made for the purpose. Assuming you don't have one, you can use a normal screwdriver on one side, but be very careful that it doesn't slip off and whack into the spinning head drum. If it does, you'll probably break off a video head.

The two guides interact with each other and also with the track at the bottom of the drum. If the center of the waveform dips while the two ends do not, at least one of the guides is too low. Try to get the waveform as flat as possible. If there's some dipping at the ends, that's okay as long as it's not severe. Video recording uses FM, so no quality of signal is lost as long as the waveform stays above a certain threshold. It'll wobble around slightly, because tape motion isn't perfect and we're dealing with insanely tight tolerances here.

As you adjust the guides, listen to the buzz of the heads hitting the tape. When the tape rides too low, the buzz increases quite a bit. The correct setting will be where it just starts to increase and sounds as stable as possible.

Unless the heads are badly worn, most of the adjustment issues will be at the start and end of the envelope, not in the middle. In particular, the entry guide side of the drum tends to have problems. To get a nice, close-up view of that transition, use delayed sweep to magnify just that portion of the waveform. Be sure to zoom back out afterward and check the entire envelope. Don't expect perfection; some amplitude variation between heads is normal. And the transitions will wobble up and down slightly. See Figure 14-9.

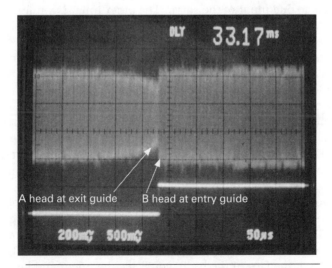

FIGURE 14-9   Magnified view of head transition

Adjusting those guides is an art form that takes a lot of experience to do well. Don't be surprised if you wind up going back and forth from guide to guide. The first few times I did this procedure were very frustrating. Stay with it, and you'll get the hang of it. If nothing works, and especially if the center of the envelope dips when the ends are correct, and you can't adjust your way out of that problem, the heads are probably worn out.

Digital Recorders

The ideas are the same, but the digital signal looks very different. You probably won't even find it in most camcorders. Your best bet is to take a tape from when the camera was new and use it as a reference, unless you have another digital camcorder whose alignment you trust. The tiny mechanisms in MiniDV camcorders bend easily, so check that the cassette is seated properly and the tension arm isn't bent. If necessary, carefully adjust the guides, watching for picture breakup instead of snow. Take note of the guide settings where it occurs and set them in the middle of the range. Use your ears to be sure the tape isn't riding too low. Also, if it's too high, the buzz will get much quieter. Find that magic spot, and you should have a good picture. Digitals use error correction and are pretty tolerant of data loss, so some misalignment gets masked, even though the drum and tape are smaller and the tolerances are tighter. I've lined up a few digital camcorders using this technique, and they all work fine and interchange with each other well, even in LP mode.

Servo Problems

Helical recorders control both head and capstan rotation with servos. Mounted on each motor are *PG coils*, which are pulse generators. The coils pick up a field from a magnet in the rotating section, generating a pulse each time the magnet goes by. This is fed to one input of the servo. The other input is a reference signal. The servo's job is to time-align the two by adjusting the motor speed and phase to match the reference signal's timing.

If a servo isn't working, the machine will mute, with no picture or sound on playback. You can verify a servo problem by scoping the RF envelope. If it has wild amplitude swings running through it randomly, the heads are not synced to the tracks because one of the servos isn't working.

Take a look at the tape reels. If they're moving fast even though the tape was recorded at slow speed, the capstan servo is out. If tape speed seems okay, look at the head drum rotation under a standard fluorescent lamp (not the spiral type). You should see a slowly creeping, almost stable pattern with a normal rate of rotation. Listen to it too. If it's wildly off, the head servo is not working.

Servo operation is complex, but it's been reduced to a couple of chips. Many servo problems are caused by loss of signal from the PG coils or one of the reference signals. The one most likely to be troublesome is the control track pulse from the audio/control head. The control track is recorded along the bottom edge of the tape. If you look at the face of the head, you'll see the upper and lower recording surfaces. Make sure the lower one is clean. Sometimes wear, misalignment or low tape tension will cause the lower surface to lose contact with the tape, wiping out playback of the control track or making it intermittent. The machine will go in and out of servo lock, showing a picture and then muting repeatedly, especially near the beginning of a tape, when tension is the lowest. (A worn or stretched tape can also cause this symptom.) Use a dry swab to press on the tape's lower edge where it meets the head while a tape is playing. If the servo suddenly locks in, there's your problem.

One of the screws on the head mounting will control the *zenith*, or forward-backward tilt. You should be able to deduce which one it is. Turn it to pull the top back a little, away from the tape. That'll put more pressure on the bottom and may restore servo operation. If you pull it back too far, the linear (non–hi-fi) audio track won't play, because the upper surface will lose tape contact.

If you really want to delve into the servos further, scope the PG coil pulses and follow their cables back to the servo circuits. Beyond that and some poking around with the scope, it's really not worth what it would take to hunt down obscure problems, given the value of the machines these days.

Color Problems

In analog machines, tape path misalignment can cause color distortion at the top or bottom of the picture by twisting or slightly stretching the tape, altering its timing characteristics as the heads read the signal. This usually occurs on the left side of the drum. You should be able to recognize such dimensional distortion visually and adjust the guide or correct the tension arm. Be sure to check the envelope after changing the tape path. You can do a quick-and-dirty check with the tracking control, watching for a reasonably even appearance of snow across the entire picture as you set the tracking away from its optimal point.

Color and luminance are processed separately in an analog video recorder and then recombined upon playback. The color subcarrier, 3.579545 MHz in the United States, is converted down to a much lower frequency before being fed to the video heads. On playback, it goes through a rather convoluted system of two PLLs that corrects the timing errors before it's mixed back into the monochrome part of the signal. If the machine plays back only in black and white, first be sure the tape you're using has properly recorded color on it. Then look for the 3.579545-MHz crystal on the board. Sometimes it's labeled 3.58. Make sure it's oscillating. You should see another couple of crystals near that one, so check their operation too. One of them is used only while recording. If you see no signal on it, put the machine into record, using another tape to avoid wiping out your test tape, and see if that crystal runs.

Audio Problems

The linear, non–hi-fi track is recorded along the top edge of the tape exactly the same way as in any audio tape recorder. The hi-fi tracks use FM subcarriers and are recorded right on top of the video tracks by an extra set of heads in a VHS machine. In Beta and 8 mm formats, the signal is mixed with the video carrier and recorded by the same heads that lay down the video tracks. In MiniDV camcorders, the audio is encoded into the bitstream along with the video data and recorded by the same heads as well.

Loss or breaking up of hi-fi audio tracks is usually due to dirty heads or mistracking. Significant mistracking can occur around the vertical sync area of the signal, off the screen. Usually that'll cause vertical jitter, but sometimes it's placed just far enough from the vertical sync to avoid affecting the image. It'll still make a mess out of the hi-fi audio, though. Check the RF envelope. If it looks okay, try cleaning the video heads again; one of the hi-fi audio heads may still be dirty.

Dropping out of MiniDV digital audio usually means misalignment or dirt in the tape path. The format covers missing video blocks pretty well with blocks from the last image, so it may be masking loss of data, but the audio will still mute.

Video Projectors

LCD and DLP video projectors are quite popular, and each type has its characteristic failures. Let's look at projectors and how to work on them.

How They Work

Front projectors and modern, non-CRT rear projection TVs use the same technologies, with a bright lamp, a "microdisplay" device that forms the image, and a series of lenses to magnify the results. In LCD projectors, there are three small LCD panels, one for each primary color of red, green and blue. Light from a very high-intensity arc lamp is filtered to remove ultraviolet energy and then split into three beams, with color filters for each color. Each beam illuminates its own panel. The resulting three images are recombined with a prism and focused on the screen with the projection lens.

DLP projectors have no LCD panels. Instead, they use a special chip called a *Digital Light Processor*, invented and manufactured by the Texas Instruments Corporation. On the surface of the DLP chip is a matrix array of microscopic mirrors, each separately addressable. Depending on the resolution, there may be hundreds of thousands to a few million mirrors. Feeding power to a mirror makes it flex, deflecting the light at an angle and reducing the amount reflected straight toward the lens. The result is a projected video image of high contrast.

That'd be all there is to it, except for one small detail: color. DLP chips are expensive, and they require a fair amount of circuitry to drive them. Some very pricey, professional-level projectors have three DLPs, combining their outputs like LCD units do. The home units you're likely to service, though, have only one chip and accomplish color projection by rapidly flashing the three color images in sequence with a high-speed rotating color wheel of red, green and blue segments between the lamp and the DLP chip. Your visual system, which can't keep up with anything coming at it that quickly, combines them into one full-color image. DLPs can move their mirrors much faster than is required for normal video rates, so it's possible for them to flash two or even three complete sets of tri-color images in the time span of one frame. When the specs say the unit has a 2X or a 3X color wheel, each frame of video is being flashed at that rate, compared to a normal video frame, with three flashes of color each time. So, a 3X-rate projector flashes nine images in the time it would take a CRT to scan one frame.

The increased rate helps diminish the "rainbow effect," an annoying consequence of the single-chip, frame-sequential color projection method that occurs when the viewer's eyes move. Especially in darker images with bright points of light, like night scenes with streetlamps, the visual trail left by the bright spot can break up into its component colors as the eyes change position, because each color frame strikes the retina in a slightly different spot, so they don't blend together. Some people find the effect very distracting, so manufacturers keep speeding up the color wheels and frame rates to minimize the time between projection of the different colors, keeping them closer together in the moving eyes of the viewer.

What Can Go Wrong

There's lots to malfunction here. The most troublesome elements in a projector are the very expensive lamp and the circuitry powering it. The brightness required is so extreme that only a high-pressure mercury vapor arc lamp will do the job. Operating an arc lamp is not as simple as just applying power. First, it has to be "struck," or started, by applying a fairly high voltage until conduction across the arc is achieved. Then, once current starts flowing, the mercury inside vaporizes and makes the lamp conduct much more readily, with lower resistance. The voltage must then be reduced to typically less than 100 volts. The circuitry driving the lamp is called the *ballast*, though it's much more complicated than the simple ballast that starts an old-fashioned fluorescent lamp.

Wait, there's more. Arc lamps exhibit some pretty odd behaviors. As they age, they tend to develop bad spots on their electrodes, increasing the resistance of the most direct path across the arc. Because the vapor conducts, other paths arise, and the arc can jump around, causing flickering of the light. To combat this annoying malady, the ballast may adjust the operating voltage or add pulses to keep the lamp at its best. Even with all this effort, lamps go bad, they go dim, they fail to strike, and now and then they explode violently.

The second most failure-prone part differs between LCDs and DLPs. In LCDs, the *polarizers*, sheets of plastic film in front of each panel, get burned by the residual ultraviolet output of the lamp, even after its light has passed through the ultraviolet filter. The blue polarizer, in particular, tends to burn, resulting in a yellowed image or splotches of yellow.

In DLPs, the color wheel, whizzing around so fast, often experiences motor failure or catastrophic mechanical failure. Because of lamp heat, color wheels are made of glass, not plastic, and some are assembled with nothing more than glue! In time, the glue degrades, also from lamp heat and ultraviolet, and the delicate red, blue and green segments fly off, smashing against the inside of the projector's case and shattering into a million pieces.

DLPs also use a light guidance arrangement quite different from that in LCDs. LCD panels are considerably larger than DLP chips, and the lamp's output is spread over enough area to illuminate them fully. That's easier than the DLP scheme, in which the light has to be formed into a small beam. To do that, DLPs use a mirror tunnel, sometimes called a *light tunnel*, made of four mirrors arranged to form a rectangular channel. Like the color wheels, the mirrors may be glued together, and the glue can fail, collapsing the tunnel.

Cooling the lamp is a critical function in all projectors, so they all have fans blowing air through the lamp housings. Most projectors have multiple fans. LCD units usually have one just to cool the panels, because, as blocking elements, they absorb a lot of heat from the lamp. DLPs, which reflect light instead of absorbing it, don't overheat their imaging chips, but they sometimes use fans to cool the rest of the optical chain, along with the lamp fan. Some projectors have power supply fans as well.

A failed fan, especially if it's the lamp fan, will cause the projector to overheat and shut down. It takes a few minutes for the thermal sensor to heat up enough to cause shutdown, so the projector will run for a short time before it dies.

Most projectors have dust filters on the lamp housings, and they get clogged with room dust to the point that airflow is severely restricted, triggering an overheat shutdown.

And, of course, projectors suffer from the usual power supply issues, especially bad capacitors. The units are remote-controlled, so at least part of the power supply runs all the time, even in standby. After a few years, a cap or two is shot, and the projector stops turning on.

Is It Worth It?

An expired lamp might seem like an obviously worthwhile repair, but the lamps cost so much—from around $100 to more than $400—that you must consider whether the rest of the projector will survive long enough to use up a new lamp. Those hot lamps put tremendous stress on the other optical components, and many projectors are designed to last about as long as one lamp. It's no fun to spend two-thirds the cost of a new projector for a replacement lamp, only to have a polarizer or a color wheel go bad 100 hours later. Some expensive, pro-level projectors are built to last through several lamp replacements, but the relatively inexpensive home units are not.

Burned LCD polarizers are pretty much a dead end unless you can scare up a parts unit. It's almost always the blue polarizer that goes, so a unit old enough to be cut up for parts probably has the same bad polarizer and will be of no use. Manufacturers don't sell the polarizers separately; they want you to buy the entire light engine (optical system) or replace the projector. Believe me, you do not want to spend what a new light engine would cost.

Color wheel costs vary greatly among manufacturers. The wheels for some rear-projection DLP TVs can be had for $50, while those for some front projectors cost an eye-popping $500. There's no basic difference between the parts; it's all a matter of marketing and volume. Parts for TVs, including lamps, are generally lower than those for front projectors.

Light tunnels can often be repaired for nothing with some epoxy and a steady hand.

The Dangers Within

The lamp is not your friend! If you look directly into it while it's running, even momentarily, I hope you like dogs, because you're probably going to need one. The brightness is higher than anything the human eye can withstand. There's a fair amount of ultraviolet, too, which is very damaging, even after most of it is absorbed by the lamp housing's UV filter. Seriously, don't ever look directly into a running projector!

The lamp gets hot enough to burn you badly, too. After it's been operating, let it cool down quite awhile before going near it. When it's hot, the glass is more fragile as well. The actual lamp envelope is only about the size of two pencil erasers, but it's under tremendous pressure. I've seen one explode, and it's not pretty. They go off like a shot, and a fine mist of glass particles tinged with mercury gets ejected from the projector's fan vent; you wouldn't want your eyes in the vicinity.

The voltages used to strike and drive the lamp are hazardous. Figure around 1 KV for striking and 80–100 volts during normal operation. Don't go poking around with your scope in the lamp supply (ballast), especially while the lamp is striking.

A DLP's rotating color wheel could cut you if you contact its outer edge while it's spinning, but it's more likely you'd destroy it.

How to Fix One

See Figure 14-10 for a view of the optical path, or light engine, in a typical DLP projector. Most problems are found there or very nearby. The three most common failures are no operation at all, no lamp strike, and overheating with subsequent shutdown. If the unit won't turn on at all, suspect the usual power supply issues. Projectors and TVs spend virtually their entire existences plugged in, waiting for a remote-control signal, so bulging power supply capacitors are pretty much a foregone conclusion eventually.

One difficulty in servicing projectors is that restarting a hot lamp damages it badly. So, once you turn the unit on, you don't want to turn it back off, take a few measurements or check a few parts, and then fire it right back up again. Always let the lamp cool before restriking it. That can take a half-hour or so.

Lamp Problems

If the projector powers up but blinks a warning light on the control panel, the lamp or its ballast may be bad. A sensing circuit checks for current draw through the lamp; that's how the thing knows the lamp has struck and it's time to reduce the striking voltage to its normal running level. If the lamp won't strike, the warning light is as far as it'll go.

You can't check the lamp for continuity with a meter; at room temperature, the bulb is an open circuit until it has 1 KV or so applied across it. Assess the lamp's condition visually. Being certain it is cool, remove the lamp. Unless the bulb itself is completely enclosed by its housing, it's a good idea to wear goggles, just in case you bump it or drop a tool on it, because it could explode in your face.

Some TVs use bare lamps, with nothing enclosing them. That little stalk protruding from the front is the end of the actual bulb, and it's fragile! At the bottom of it, nearest the back of the reflector, is the high-pressure envelope. The lamps in most projectors

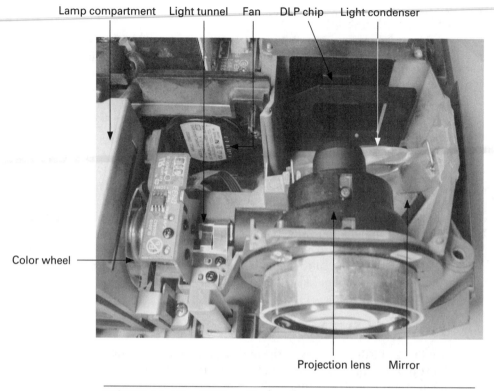

Lamp compartment   Light tunnel   Fan   DLP chip   Light condenser

Color wheel

Projection lens   Mirror

FIGURE 14-10   DLP optical path

are enclosed in a housing with a UV filter in front, so they're a little safer, but you could still get showered with glass blown out the sides of the housing, through the dust filters.

It's important *never* to touch the bare lamp or the UV filter, because skin oils will make them crack when they get up to operating temperature. A cracked UV filter may leak UV radiation, and a cracked bulb...well, you know what will happen. Kaboom!

Look at the envelope head-on, slightly from the side. You're looking down the focal point of a parabolic reflector, so it's hard to see the arc gap, but if you look a bit off-axis, you can see it. It'll look greatly magnified by the reflector, and that's helpful. If the envelope's glass looks clear and clean, the lamp doesn't have lots of hours on it and is probably good. If it looks charred, it's an old lamp and may be shot. Examine it carefully, and you can also see the actual electrodes and their condition.

Almost all projectors have a time counter in their menus telling you how many hours are on the lamp. Without a working lamp, of course, you can't see it!

Ballast Problems

The ballast is really a pretty fancy power supply of its own. It has to supply the high striking voltage and then the lower operating voltage. Operation can require a few amps, for around 150–250 watts of lamp power. Output transistors supply it, and they can pop. Also, a fair amount of heat is generated in the output stages of some ballasts, so check the circuit board for burn marks and degraded solder joints.

Don't try scoping the ballast's output stages. It's far safer and easier to disconnect power and pull and check the output transistors. There may be an onboard fuse, too. Be sure to discharge any large electrolytics before desoldering anything.

Look for optoisolators in the path between the lamp and the ballast. The output of one of them will change state when the lamp successfully strikes, relaying the information that it's time to lower the voltage. If you can't find that signal, the lamp is not being struck, or it's bad.

Overheating Problems

Overheat shutdowns don't happen instantly; it takes a little while for the heat to build up. If the projector runs for five or ten minutes and then quits, it's probably overheating. A lamp very near the end of its life can run excessively hot and cause this condition, but most of the time it's due to lack of airflow over the bulb. Check first for blocked dust filters. Some projectors, especially LCDs, have them at the air intakes for both the lamp and the LCD assembly. Blockage of either can trip the shutdown. DLPs usually have filters right on the lamp housing. Some units are designed to be dust-resistant and have no filters. They rarely clog up, but check the air channels at the lamp housing just to be sure nothing has gotten in and blocked those.

There's a thermal sensor over the lamp. If the fan doesn't do its job, the lamp overheats and the unit shuts down. Never defeat this sensor, even just for a repair test. Even if you don't cause a fire, you'll probably destroy the lamp, and it could get hot enough to burst.

The fans themselves can fail, but sometimes the problem is the circuitry powering them, especially in units with variable-speed fans. Many projectors have normal and economy settings for lamp brightness, with greatly increased lamp life at the lower setting. To reduce noise, they slow the fans down in the economy mode, using power transistors to lower the voltage. Over time, a transistor may fail, and the fan will go dead. The fans should always be turning when the projector is on. If one of them isn't spinning, disconnect power and try turning it by hand. If it's not gummed up with dirt or dried-out lubrication, it should move freely.

The fans are the same types used in desktop computers, and they usually run on either 5 or 12 volts when at full speed. Check to see if any voltage is getting to them. It may be less than the fan's specified voltage, but it won't be a fraction of a volt, and it'll never be zero if things are working. Expect at least a couple of volts. Don't use circuit ground for this test; check directly across the fan's positive and negative leads. If a fan shows voltage across its leads but is not moving, the fan is bad. You can probably find a compatible replacement fan from a computer supply house. Just be sure the new one moves at least as much air.

If there's no voltage, the driving circuit is out. Trace the fan's wires back to the board and look for a small power transistor. Scope to see if it's getting power supply voltage and if that is getting fed to the fan. The transistor may be between the fan's negative terminal and ground, with supply voltage going to the fan's positive terminal.

If you see the same voltage on the fan's negative terminal, the transistor isn't pulling current to ground and is probably open. That's why you can't use circuit ground when testing the fan; you need to be sure there's actually a voltage difference across it.

Light Tunnel Problems

If the light tunnel collapses, the projector will run but you'll see a darkened area along an edge or even large portions of the image blacked out. Gluing it back together is a chore, but the price is right! Carefully disassemble the optical path and remove the tunnel's mirrors. The shape of the tunnel corresponds to the aspect ratio of the imaging chip. You should be able to deduce how it went together from the position in which you find the pieces, and from the glue remnants. Clean off the old adhesive and glue the tunnel back together with epoxy; instant glue will not survive the lamp's heat. Use the good stuff that takes awhile to dry, not the 5-minute quick-set type. If you can find high-temperature epoxy, that's your best bet. Let it dry for a night or two and then reassemble.

Color Wheel Problems

The DLP's color wheel is driven by a small motor, with a position sensor to tell the video circuitry when to flash the correct image for whatever color is in front of the lamp. Most of them use a Hall-effect sensor that picks up a magnetic field from a magnet embedded into the assembly's rotor, generating a pulse for each revolution of the wheel. Without that signal, the unit's microprocessor assumes the wheel isn't turning properly, or at all, and stops operation, shutting down the lamp.

Treat the color wheel gently. It's delicate, and breaking it means the end of the projector unless a petal has come off intact, in which case you can glue it back on. When the unit starts up, see if the wheel is turning. It should spin very fast. If it seems sluggish or isn't moving at all, unplug the projector and gently turn the wheel by hand. If it doesn't turn freely, the motor's lube may have dried out. Often, dried lube will cause a horrible screeching sound from the wheel, too. The normal remedy is replacement, but the cost can be ridiculous.

I've saved a few screeching color wheels for nothing, though, and the fix has lasted a long time. Get some silicone lubricant spray. Don't use anything else, like oil or other kinds of sprays. Remove the wheel and spray into the motor bearings, saturating the motor as best you can. You'll probably be spraying under the color disc itself into the top of the motor. Let it soak in for a few minutes, and then dump out what's left into a tissue. Very gently clean excess lubricant off both sides of the color disc with an alcohol-soaked swab. Don't put pressure on that disc or you may snap off the glass petals. Give the motor a few careful spins and then reassemble the unit. Sometimes it works.

Video Processing Problems

The circuitry driving the LCD panels or the DLP chip is dense and complex. It's all digital and not very serviceable. If you're losing a color on an LCD, or the image is flickering or freezing on a DLP, check for bad cable connections to the boards driving those imaging devices. Also look for leaking surface-mount electrolytic capacitors. The heat of a projector can really shorten their life. If there's no video at all, look for on-board voltage regulators, and scope to see if they're putting out voltage. Beyond that, you'd need a service manual and some serious, high-speed digital test gear to make much sense out of the signal processing. These kinds of problems aren't common, though. The vast majority of projector failures can be traced to the power supply, the ballast, the lamp and the optical components.

Blobs in the Image

These are caused by dust on the LCD panels or the DLP chip. A little carefully applied compressed air will take care of them. You may have to open a housing or remove the lens to gain access. Be careful not to spray the air too close to the imager, in case some of the cold propellant winds up on the optical surface, where it could mar the transparency. The actual mirrored surface of a DLP chip is behind a sealed window, so there's no chance of harming the mirrors. LCD panels and their polarizers are more easily damaged.

Have at It!

I hope you've enjoyed our extended romp through consumer electronics and its repair. Armed with the techniques presented in this book, you should be ready to dig in and have some fun! Just always remember to put safety first, and keep in mind that, as with any skill, expertise develops over time. If you wreck a product while trying to fix it, toss its remains on the parts pile and chalk up the loss to a valuable lesson learned. We all lose a few along the way. The more you practice, the better you'll get.

Ready to rock? Scope on? Soldering iron warmed up? Ignition sequence start...T minus 3, 2, 1...blastoff!! Have a grand adventure!

Glossary

Here are some of the common terms you'll encounter when working on electronic devices. There are plenty more, but you'll see these often.

active elements   The semiconductors at the heart of a *circuit stage* which perform amplification, switching or other signal processing functions defining the stage's operation. Transistors and IC chips are active elements. Before solid-state electronics, the active element was the *vacuum tube*.

ADC   Analog-to-digital converter. A device that converts analog signals to a digitally coded representation.

alternating current (AC)   The polarity periodically reverses direction. Reversals may occur at any rate above zero cycles per second, or *hertz*, up to billions per second.

amperes (amps)   The volume of flow of electric charges per unit of time, or *current*. This corresponds to how much electricity is flowing, independent of how hard, and is represented as *A* or sometimes *I*. In small-signal circuitry, it is usually specified in *ma*, or milliamps (thousandths of an amp). The polarity of the current will always be the same as that of the voltage, as the voltage is what drives the current through the circuit, so it has to be going in the same direction.

amplitude   The voltage strength of a signal. For audio, it represents loudness. For video, brightness.

analog   A method of representing information by varying a voltage over time in a continuous pattern resembling the information, distinctly different from the either/or, on/off states of *digital* representation.

audio   An analog signal representing sound by varying a voltage in the same pattern as the original mechanical vibrations of the sound.

audio-frequency (AF)   A signal in the frequency range of 20 Hz to 20 kHz, the approximate *spectrum* of human hearing.

ballast   The circuit used for starting a lamp. Ballasts range from very simple, as with old-fashioned fluorescent lamp tubes, to complex, as with the high-pressure arc lamps used in video projectors.

bias The DC current or voltage supplied to the input of an amplifying element to keep it in its conducting state during the required portion of the signal waveform.

bipolar A circuit with both positive and negative power supply polarities relative to ground.
Also, the internal construction of an NPN or PNP transistor.

bus A large connection point used to distribute power or provide ground to multiple areas of a circuit. In digital devices, the shared connections used to distribute data to multiple points.

bypass capacitor A capacitor connected to shunt a signal's AC component to ground without having an effect on the signal's DC component, if there is one.

capacitance Specified in farads, capacitance is a measure of how much charge can be stored on the two plates of a *capacitor* for a given applied voltage. The more charge stored, the more current will pass back out of the capacitor when it is discharged into a circuit.
Numerous factors affect capacitance, including surface area of the capacitor's plates, their distance apart and what separates them. Most capacitors are specified in microfarads (μf, or millionths of a farad) or picofarads (pf, or trillionths of a farad). In European gear, capacitors specified in nanofarads (nf, or billionths of a farad) are sometimes used.

carrier A signal upon which another, lower-frequency signal is imposed. In radio, the high-frequency signal emanating from the transmitter, on which is imposed amplitude or frequency *modulation* of audio information.

CCFT Cold-cathode fluorescent tube. The type of lamp used in LCD backlights. CCFTs have no filaments. They are driven with high-frequency, high-voltage pulses applied across the length of the tube, and do not require a *ballast*.

chopper The transistor used in a switching power supply to convert incoming power into high-frequency pulses that will be fed to the conversion transformer.

clipping The state of nonlinearity in an amplifier that occurs when an input signal attempts to drive its output to voltages greater than those provided by the power supply. The amplifier's output signal can't swing as far as the input signal demands, so it stops at its limits and clips off the tops and bottoms of the waveform, causing severe distortion. See *linear*.

CMOS Complementary metal oxide semiconductor. This type of integrated circuit construction is widely used because of its wide voltage tolerance and low-power operation. Once limited to small-signal circuits, CMOS devices now handle serious voltage and current, and are found in MOSFET power transistors and output modules.

Some CMOS devices are easily damaged by static electricity because it can punch holes through the oxide layer, causing an internal short circuit.

complementary   Opposite in phase. Also, an amplifier design with separate transistors operating on opposite halves of the signal waveform. A true complementary design uses PNP and NPN (opposite-polarity) transistors. One using the same polarity transistors for both waveform halves is called *quasi-complementary*. Both schemes are used frequently in audio power amplifiers, with the two transistors shown one above the other on the schematic diagram. Complementary amplifiers are called Class AB.

coupling   The passing of a signal from one circuit stage to another. Typically, the function of a capacitor when it transfers the AC component of a signal from one stage to the next, blocking the DC component.

CRT   Cathode ray tube. A picture tube, as used in older TVs and computer monitors.

current   The flow of electric charges. Though it is semantically incorrect to describe the "flow of current," since that is the "flow of the flow of charges," most of us in electronics think of current as "the quantity of electricity," and describing current as flowing is common and considered acceptable by most practitioners. See *amperes*.

curse words   Utterances sometimes emitted by technicians, immediately following the appearance of *magic smoke*.

cutoff   The state of a transistor when it is fully turned off, allowing no current to pass through it.

cutoff frequency   The signal frequency at which a transistor's gain has decreased to 1, or unity gain.

DAC   Digital-to-analog converter, sometimes pronounced "dack." A device that converts digital codes to analog signals.

damper diode   A diode connected across an inductor to short out reverse-polarity current, generated when the inductor's magnetic field collapses, which could damage other components. Damper diodes are connected cathode to positive, so they have no effect on the current applied to the inductor. They are most often used with relay coils and motors. Some relays include the diode inside.

Darlington pair   An arrangement of transistors with the output of one directly feeding the base of the next, to multiply their gains. With NPN transistors, the collectors are tied together, and the emitter of the first transistor is connected to the base of the second. Some Darlington pairs are built into a single transistor package, providing gain unattainable with just one normal transistor. Darlington pairs are often found in audio power amplifiers.

DC-DC converter   A small *inverter* used to step up a DC voltage to a somewhat higher one. Typically used to generate 15–30 volts in devices operating from 3–6 volts.

demodulation   The extraction of information from a signal. Also called *detection*. See *modulation*.

digital   A method of representing information by measuring it in a series of *samples,* which are then encoded into a pattern of voltages having only two states, on or off, representing the binary numbers 1 and 0.

direct current (DC)   The current moves only in one direction. Current moves from negative (an excess of electron charge) to positive (a dearth of it). Most modern gear uses negative ground, so the current is really moving up from the circuit ground point, through the components and then into the positive power supply connection. It's convenient to think of ground as the sinkhole into which everything is flowing, though, and that's perfectly acceptable regardless of polarity. Conceptually, the actual direction of electron charge flow is irrelevant. (In physical terms, the direction of flow is quite important, however, as you'll disastrously discover should you try to run a product from a power supply with the polarity reversed.)

discrete   Made of individual parts, as opposed to integrated circuits. An audio output stage with separate power transistors, instead of an integrated module, is a discrete output stage.

distortion   The unfaithful reproduction of a signal by an amplifier, in which the output does not accurately mimic the input. See *clipping* for one example.

DLP   Digital Light Processor. The trade name of a Texas Instruments video projection technology employing microscopic mirrors on a chip that can be flexed with applied voltage, aiming reflected light toward or away from a lens. Practical DLPs may have millions of mirrors.

DSP   Digital signal processing. The term is applied to both the process and the specialized chips performing it, particularly in the filtering or enhancement of audio signals.

DUT   Device under test. The circuit or component on which you're taking measurements.

duty cycle   The percentage of time that one cycle of a signal's waveform spends in its "on" state. This is usually used to specify the percentage of time a square wave spends turned on. A wave with equal on and off times has a 50-percent duty cycle.

emitter follower   A current amplifier, or *buffer*, in which the output is taken from the transistor's emitter. Emitter followers offer only current gain, not voltage gain, and they do not invert the signal. The output is a replica of the input, except that it has higher current driving capability, so it can feed a greater load.

envelope   The overall contour of a waveform, viewed over many cycles.

ESR   Equivalent series resistance. The internal resistance of a capacitor that limits how fast it can be charged and discharged. Some ESR is normal, but electrolytic capacitors often develop increased ESR as they age, eventually causing circuit malfunctions.

eye pattern   The distinctive *envelope* of the output signal from an optical disc player's laser head.

fall time   The time it takes for a waveform to drop from 90 to 10 percent of its maximum value.

feedback   A signal sent from a circuit's output back to its input. The path the signal takes may involve intermediary components, and is sometimes called a *feedback loop*.

filter capacitor   A large bypass capacitor used to smooth the output voltage of a power supply.

frequency   Specified in hertz (Hz), how many times per second an event occurs. See *alternating current*.

gain   The ratio of the strengths of the input and output signals of an amplifying circuit. *Voltage gain* indicates that the circuit creates a replica of the signal, but with a bigger voltage swing from peak to peak. *Current gain* means that the circuit's output signal may have the same voltage swing as its input but is able to supply more current, as would be required, for instance, when driving a speaker.
   Also, the ratio of base-to-emitter current to the resulting collector-to-emitter current in a transistor. For example, if 2 milliamps of base current will cause 100 milliamps of collector current, the transistor has a gain of 50. A transistor's gain value is engineered into its manufacture, and it decreases with signal frequency.

giga   Billions.

ground loop   A condition in which the ground terminals of two connected pieces of equipment aren't at the same voltage level, so current passes between them, causing AC line noise to corrupt signals. Most ground loop problems occur with analog audio and video gear, resulting in audio hum or bars moving up the screen of a video display. In digital systems, ground loops may result in corruption severe enough to cause dropouts or complete loss of data transfer.

Hall-effect sensor   A semiconductor *transducer* whose passage of current is affected by nearby magnetic fields. These sensors are commonly used to detect rotation and position in servo-controlled motor systems, as in VCRs and camcorders, especially in their capstan motor assemblies. Unlike inductive sensors, Hall-effect devices will respond to the steady field from a permanent magnet not moving relative to the sensor.

harmonically related   Two waves that are related in frequency by a whole-number ratio, such as 2:1, 3:1, and so on. They are considered related because their cycles will have starting and stopping points that correspond in a regular, repeating pattern.

harmonics   Parts of a wave whose energies are at frequencies that are a whole-number multiple of the fundamental frequency of the complete wave. A square wave, for example, contains energy at all the odd harmonics (3rd, 5th, 7th, and so on) but none at the even harmonics.

heatsink   The metal structure that carries heat from an attached component, radiating some of the heat into the air and keeping the component cooler than it would be on its own. Heatsinks usually have fins to maximize surface area for greater cooling. Sometimes the metal chassis of the product can serve as a heatsink, with power-handling components bolted to it.

impedance   Usually specified in ohms, like *resistance*, impedance is a complex quantity describing the opposition to AC currents. Along with resistance, it includes the effects of capacitance and inductance, known as *reactance*. Think of it as resistance, but for AC. Unlike pure resistance, the impedance of a circuit varies with the frequency of the applied signal. The impedance of a capacitor, called its *capacitive reactance*, drops as signal frequency rises. The impedance of an inductor does just the opposite.

inductance   Specified in *henries*, inductance describes the voltage that builds up in a coil of wire when a changing current passes through it. The effect arises from the magnetic field the current creates and how that field impinges on the coil, opposing its passage of current. It is determined by physical factors inherent in the coil, including the number of turns of wire, whether it is wrapped around an iron core, and other conditions. In small-signal circuits, inductance is specified in millihenries (mh, or thousandths of a henry) or microhenries (μh, or millionths of a henry).

input stage   The low-level circuitry used to amplify small signals from *transducers*.

integrated   Having many components formed on the same *substrate*, or base layer, as in an integrated circuit or module.

inversion   A signal stage flips the signal upside down, so the output voltage drops as the input voltage rises, and vice versa. Amplifiers providing *voltage gain* often perform inversion as well, and are known as *inverting amplifiers*.

This term is unrelated to the use of the word *inverter* to describe a step-up voltage converter of the sort used to drive fluorescent lamps in LCD screens.

kilo   Thousands.

land   The copper area on a circuit board where components are connected, distinct from the *traces* connecting such areas. Some techs call all conductive elements on circuit boards traces.

latching or latch-up   The destructive condition that may occur when an integrated circuit's signal input voltage exceeds the power supply voltage, causing reverse current through the *substrate*.

LCD   Liquid crystal display. An electronic light valve in which the molecules of a special liquid with crystalline properties can be twisted to alter the polarization of light passing through them, causing darkening when a fixed polarizer placed in line no longer matches the polarization of the altered area. Practical LCDs may have millions of individual light valve cells.

lead dress   The placement and securing of wires and cables in a device.

line voltage   The voltage of the AC line as it comes from the wall socket.

linear   A graph of a linear circuit's input on one axis and its output on the other creates a straight line, indicating that the circuit is faithfully reproducing the signal, without distortion. Especially in high-fidelity amplifiers, linearity is an important design goal. Some circuits, such as radio mixers, are deliberately nonlinear, so that two signals will influence each other, creating new signals.
   Also, a general term meaning analog, as opposed to digital.

load   How much *resistance* (DC) or *impedance* (AC) a device presents to the circuit driving (powering) it. The lower the resistance or impedance, the greater the load, because more current will be needed to drive the device. Most speakers, for instance, offer an 8-ohm load to the amplifiers driving them.

LSI   Large-scale integration. A large-scale integrated circuit chip may have thousands or millions of elements.

magic smoke   The essence that actually runs electronics. Let the magic smoke out, and the circuit won't work anymore. All pro techs know this because they've made it happen, usually to something expensive. See *curse words*.

matrixed   Addressable in an X-Y grid, as in an LCD or a keypad.

mega   Millions.

micro   Millionths.

milli   Thousandths.

modulation   The altering of a signal to impose information on it. In amplitude modulation, or AM, a *carrier's* strength changes with audio information. In frequency modulation, or FM, a carrier's frequency is shifted higher and lower to convey the information. Many other forms of modulation are used, especially in digital systems. See *demodulation*.

multiplexed   A circuit in which information to or from separate areas is carried on common lines through time sequencing, or scanning. Computer keyboards and remote control keypads are multiplexed, with a microprocessor scanning a small set of lines in an X-Y style grid, looking for intersecting connections indicating which button has been pressed. LCDs are also scanned in such a grid. The technique permits a small number of lines to serve a much larger number of switches or screen pixels.

Also, a *subcarrier* system used in analog FM radio to encode stereo onto a single carrier.

mu-metal   A special metallic alloy, particularly effective at blocking electromagnetic fields, used in shields over sensitive components and input stages.

nano   Billionths.

NC   No connection. Mostly used to denote IC pins that either have no connection to the inside of the chip or are not connected to anything in the rest of the circuit.

negative feedback   A signal fed from a circuit's output back to its input, but upside-down, or 180 degrees out of phase, for controlling distortion or motion, or any other corrective action.

Ohm's Law   Georg Simon Ohm's crucial contribution to the electrical art, stating that voltage (E) equals current (I) times resistance (R). Knowing any two of the variables makes it easy to solve for the third one. Thus, $I = E / R$ and $R = E / I$. Using the more common terms of volts, amps and resistance, $V = A \times R$, $A = V / R$ and $R = V / A$.

open circuit or open   A disabled circuit path preventing passage of current required for proper circuit operation, as through an *open* component. A switch in the "off" position is also said to be open, as no current passes through it.

output stage   The high-level circuitry used to deliver sufficient power to a final transducer or device such as a speaker, motor, transmitting antenna or backlight lamp.

parallel   Components are connected together across the power source, so current goes to each one, independent of the others. In a parallel circuit, the voltage is the same to each element, and the current that passes through each one is proportional to its resistance.

passive elements   The non-amplifying components supporting the operation of a circuit stage's *active elements*. Resistors, capacitors and inductors are passive elements.

PC board or PCB   Printed circuit board.

phase   The relative position in time of two signals, expressed in degrees as a fraction of one cycle's 360-degree total. When the signals are half a cycle apart, they are 180 degrees out of phase.

photodetector   A light-sensitive component, usually a phototransistor.

pico   Trillionths.

pictorial   A drawing of a device's internal structure in physical terms, including appearance and layout of its parts.

PIV   Peak inverse voltage. The maximum voltage a diode or rectifier can withstand in the direction the part does not conduct.

pixel   A single picture element, or dot, especially in a *matrixed* display like an LCD.

polarity   The direction of current, as indicated by + (positive) and – (negative).

polarized   A component that requires a specific arrangement of positive and negative voltage applied to its terminals. For example, most electrolytic capacitors are polarized and will be destroyed if voltage is connected with the polarity backward.

power transistor   A large transistor capable of passing significant amounts of current and dissipating more than a few watts as heat. Power transistors are used as output elements in amplifiers, as motor drivers, and anywhere else their large power-handling capabilities are required.

push-pull   A common form of high-fidelity audio amplifier in which the positive and negative halves of the signal waveform are amplified separately and recombined at the output. See *complementary*.

PWM   Pulse-width modulation. The technique of varying the *duty cycle*, or on/off ratio, of a series of pulses to convey information or control the flow of power. Switching power supplies use PWM to regulate their output voltages, and PWM is also used to vary the speed of motors in some applications.

quiescent current   The current drawn by an amplifier stage when no signal is present. Quiescent current is set by the active element's *bias*.

radio-frequency (RF)   Generally from around 100 kHz to many GHz (gigahertz, or billions of cycles per second).

rail   The power supply line feeding a circuit. Though some circuits with inductors can generate voltages exceeding those of the rails powering them, most cannot, and the rail's voltage is considered the circuit's limit.

rectify   To convert AC to DC, either by chopping off one half of the waveform or directing opposite halves to the appropriate + and – terminals, so that neither output terminal changes polarity as the incoming current alternates.

resistance   Specified in ohms, resistance opposes the current moving through any conductor or circuit, limiting the total amount. It is essentially atomic friction, and the power lost to it is converted to heat. It is represented as R and by the Greek letter omega, Ω. Resistance has no polarity, just as a crimp in a hose has the same effect regardless of the direction of water flow.

resonance   The state in which a component or combination of components builds up a voltage or current when an AC signal of a specific frequency is applied, because the transit time of the energy through the circuit is such that energy peaks reflecting back and forth through the components coincide with the incoming peaks of the applied signals, reinforcing them. The electrical effect is similar to the mechanical effect that occurs in a musical instrument's string, and resonant circuits are said to be *tuned* to their resonant frequency.

ripple   The unwanted variations in the output of a DC power supply.

rise time   The time it takes for a waveform to rise from 10 to 90 percent of its maximum value.

RoHS   Restriction of Hazardous Substances. The European standard specifying lead-free soldering and other environmentally safer construction characteristics. Devices built to this standard will display RoHS on a label or stamped into the case.

sampling   The process of taking rapid measurements of analog voltages to convert them to digital data representation. See *digital*.

saturation   The state of a transistor when it is fully turned on, allowing maximum current to pass through it with minimum resistance.

sawtooth wave   An asymmetrical waveform whose shape resembles the teeth on a saw, typically used in the beam-sweeping circuitry of CRT TV sets and oscilloscopes.

schematic   A circuit diagram showing the symbols and interconnections of components, without regard to their size, appearance or physical layout in the device.

selectivity   The ability of a receiver to separate adjacent stations.

semiconductor   A component, usually made from silicon, such as a transistor, diode or IC chip. Some exotic semiconductors are made from other materials like gallium arsenide or germanium, which behave similarly to silicon in their ability to control the flow of electrons in response to a control signal.

series   Components are connected end-to-end, so current must pass through one to get to the other. A fuse, for instance, is connected in series with the circuitry it's protecting, so that all current must pass through the fuse to reach the rest of the components.

In a series circuit, the current is the same through each element, but the voltage reaching each element is reduced in proportion to the other elements' resistance. The lowering of voltage is referred to as an element's *voltage drop*.

series pass transistor   A transistor used as a variable resistance element in a linear voltage regulator to hold the output voltage to a specific value, with all of the power supply's current passing through the transistor on its way to the rest of the circuitry.

shield   A metal cover placed over sensitive circuitry to protect it from stray electrical noise. Shields are often soldered in place and must be removed for access.

short circuit or short   A low-resistance passage of current where it shouldn't exist, as from a voltage point directly to circuit ground via a *shorted* component.

shunt   A current path in parallel with another path. A resistor across another component, diverting some current around it, is said to be shunting the component or shunting the current.

signal   A changing voltage representing information.

sine wave   A waveform with a smooth, repetitive shape derived from the mathematical *sine* function. Sine waves have no *harmonics*, or energy at multiples of their frequency, so they are the purest type of signal possible.

sink   An acceptor of current that permits it to pass toward ground. See *source*.

sled   The linear track assembly on which a laser optical head rides as it scans across the disc. The sled includes the track rods and the motor and gears moving the head.

SMD or SMT   Surface-mount device or surface-mount technology. These are the tiny, leadless components that have replaced conventional through-hole parts with leads.

SMPS   Switch-mode power supply. Another name for a switching power supply.

SNR or S/N   Signal-to-noise ratio. The ratio of desired signal to residual noise in a circuit, usually expressed in decibels (db).

source   A supplier of current. See *sink*.

spectrum   A range of frequencies.

square wave   A waveform with flat tops and bottoms that switches quickly between the two states, with fast rise and fall times. Square waves can never be truly square, because it takes some time for them to switch states.

stage   The distinct section of a circuit that performs one function of signal processing. Circuit stages are organized around one or more *active elements*, with surrounding *passive elements* supporting their operation.

subcarrier   A *carrier* signal imposed on another, higher-frequency carrier. The technique is used to piggyback hidden signals onto others, as in FM stereo and analog color video signals.

substrate   The base on which an integrated circuit, transistor or potentiometer is formed. In most modern solid-state components, the substrate is a slice of silicon.

switch mode   The operation of a transistor as a switch—that is, *saturated* or *cut off*—with fast rise and fall times between the two states.

sync   The pulses in an analog video signal that align the position of a TV's display point (the electron beam's position in a CRT TV or the row and column addresses in a matrixed display like an LCD or plasma TV) with that of the originating signal source, to keep picture elements correctly placed.

time constant   The period of time it takes to charge or discharge a combination of components, such as a resistor and capacitor or an inductor and capacitor.

tough dog   A unit with a repair problem that's difficult to diagnose.

trace   The conductive copper lines on a circuit board providing connection between *lands*, or component connection points.
    Also, to follow a signal through a circuit with a measuring instrument, especially an oscilloscope, or to follow a connection visually across a circuit board or a schematic diagram.

transducer   A device that takes in one kind of information or energy and converts it to another. Microphones, tape heads, laser optical heads, speakers and phono cartridges are all transducers.

video   An analog signal representing a moving image by varying a voltage to represent the brightness and color values of scanned spots in the image. Video signals are complex and include synchronizing and color reference signals to align the receiver's circuits as necessary to interpret the signal properly and display the image.

volt-amps (VA)   This is somewhat like *watts*, but for AC power. It means "volts times amps." Because of certain effects that occur with AC, the maximum voltage peak and maximum current in each cycle may not coincide, and the amount of work that can be done will be less than when they do. The power requirements shown on the backs of AC-powered devices are often specified in VA. For repair work, it's safe to think of VA and watts as equivalent, since actual watts used will never exceed VA and may be less.

volts   The electric "pressure" or *electromotive force*. This corresponds to how hard the electrons push, and is represented as *V* or sometimes *E*. Voltage propels current through resistance, so the higher the voltage, the more current will pass through a given resistance. (See Ohm's Law). *Vcc* and *Vdd* are references to voltages applied to the pins of a semiconductor, and mean the same thing as V, except that they are always of positive polarity. Negative voltage is sometimes noted as *Vss*.

Like all other motion of mass in the universe, voltage is entirely relative; there is no such thing as "9 volts." A given voltage is meaningful only relative to some other point. Polarity is relative too; a voltage can be positive with respect to one point while being negative with respect to another! In circuits that have both positive and negative voltages relative to circuit ground, that ground point will function as positive for the negative voltage point and negative for the positive voltage point.

watts   Also called *power*, a measure of how much work can be done by the electricity. Power is equal to volts × amps. So, 100 watts could be 20 volts at 5 amps or 10 volts at 10 amps; the amount of work either of those quantities could do would be the same. 746 watts equal one *horsepower*, but if you cause 746 watts of electrical power to flow through an actual horse, you wind up with a dead horse and zero horsepower.

Watt's Law   This states that power in watts (P) equals current (I) times voltage (E). As with Ohm's Law, knowing any two of the variables makes it easy to solve for the third one. Thus, I = P / E and E = P / I. Or, in the terms of watts, amps and volts, W = A × V, A = W / V and V = W / A.

Common Circuits

Many devices are constructed from the same basic types of building blocks. Here are some widely used circuits you may find.

amplifier   One or more stages that increase a signal's voltage swing, current or both, by shaping power supply current into a larger replica of the signal.

buffer   An intermediary circuit isolating two others. In analog devices, an amplifier stage providing current gain and isolating the more sensitive, earlier amplifier stage from the ones following it. Buffers typically offer no voltage gain. In digital processing, a temporary storage area holding blocks of data, perhaps in line for conversion to analog signals.

clock oscillator   The square wave oscillator used to step, or clock, a digital circuit through its series of operations. Many microprocessors used in small products include onboard clock oscillators, requiring only a crystal or ceramic resonator to set the frequency.

crowbar   A circuit designed to detect a fault condition, usually overcurrent or excessive voltage, and deliberately blow the power supply's fuse to stop operation.

frequency synthesizer   A complex circuit used to generate a range of frequencies from the single frequency of a quartz crystal. Older frequency synthesizers used *voltage-controlled oscillators* and *phase-locked loops* to lock an analog oscillators' signals to the frequency determined by a digital controller. Many newer systems use *direct digital synthesis*, in which the required analog waveforms are generated by the digital system via a digital-to-analog converter in much the same way a CD player reconstructs the analog audio waveform from digital samples.

front end   The sensitive first stages of a radio receiver that accept energy from the antenna and amplify it in preparation for its delivery to the *mixer*. Many front ends include tuned circuits to help reject off-frequency signals, but those in frequency-synthesized receivers may not.

intermediate-frequency (IF)   The fixed-frequency, tuned amplifier stages in a radio receiver providing most of the selectivity (ability to separate stations) and sensitivity. Incoming signals from the antenna are converted to the IF by mixing them with a local oscillator in the *mixer* stage, a deliberately nonlinear circuit that causes the two signals to interfere, generating a new signal at their difference frequency that still carries the information of the original signal.

The mixing process is also called *heterodyning*, and radios that convert signals to an IF, which virtually all modern sets do, are called *superheterodyne* receivers, because the output of the heterodyne process is still higher up the frequency spectrum than the information carried by the signal.

Also, the frequency at which the tuned amplifiers operate.

inverter   A step-up switching power supply typically used to generate the high voltages required for fluorescent LCD backlights, camera flash tubes and other applications needing voltages higher than those provided by the product's primary power supply or batteries.

linear voltage regulator   A step-down regulator (one whose output is at lower voltage than its input) using a transistor element in *linear mode*, as a variable resistor that dissipates excess power as heat.

mixer   A circuit that accepts two AC signal inputs and outputs the sum and difference frequencies resulting from their interacting, or mixing, with each other in a nonlinear manner.

oscillator   A circuit that produces a steady signal of constant frequency, or variable frequency in some cases, through the use of constructive, or reinforcing ("positive"), feedback.

phase-locked loop (PLL)   A multistage circuit used to slave the frequency of an oscillator to that of a reference signal or another oscillator. By inserting digital dividers between the oscillator and the rest of the loop, the oscillator can be forced to lock to a multiple of the reference frequency, a crucial technique in frequency synthesis. PLLs are used to tune receivers, to recover color information from analog videotape and data from the laser head's signals in CD and DVD players, and in many other applications.

power supply   Any circuit or power source that provides the power the rest of the device requires. In AC-powered products, it typically converts the incoming line voltage to a steady, lower DC voltage. In battery-operated products, the batteries themselves may be considered the power supply. If intermediate circuitry steps the battery voltage up or down, that circuitry will also be considered part of the device's power supply.

preamplifier or preamp   A low-level amplifier used to boost weak signals enough that further stages can amplify them more.

servo   A circuit that slaves mechanical motion to some reference, typically used to lock the rotational speed or phase of a motor to a signal, as with a VCR's rotating video head drum, which must be kept aligned to the recorded tape tracks for playback.

switching or switch mode power supply   A power supply operating in *switch mode*, in which the incoming power is chopped into high-frequency pulses before being fed to the conversion transformer. This approach allows a small transformer to provide substantial power output, and it is very efficient when compared to a linear power supply, which processes power at the very low line frequency of 50 or 60 hertz. Switching power supplies can be used to step voltage up or down. Step-up versions are often referred to as *inverters*.

switching or switch mode voltage regulator   A voltage regulator operating in *switch mode*, in which the duty cycle of a pulse is varied to regulate the output voltage. This is similar to the operation of a switching power supply, except that the incoming power is already at a DC voltage in the neighborhood of the desired output voltage. Compared to a linear voltage regulator, a switching regulator generates far less heat and is much more efficient. Plus, it can be used to step the voltage up or down.

voltage-controlled amplifier (VCA)   An amplifier whose gain can be controlled by a DC voltage. VCAs are used in the audio sections of digital devices such as MP3 players, with the DC voltage from the microprocessor setting the headphone volume.

voltage-controlled oscillator (VCO)   An oscillator whose frequency can be controlled by a DC voltage. VCOs are used in radio tuning circuits, with the DC voltage from the digital circuits setting the frequency.

Index